T0183777

Communications in Computer and Information Science   **966**

Commenced Publication in 2007
Founding and Former Series Editors:
Phoebe Chen, Alfredo Cuzzocrea, Xiaoyong Du, Orhun Kara, Ting Liu,
Krishna M. Sivalingam, Dominik Ślęzak, and Xiaokang Yang

Editorial Board

Simone Diniz Junqueira Barbosa
Pontifical Catholic University of Rio de Janeiro (PUC-Rio),
Rio de Janeiro, Brazil
Joaquim Filipe
Polytechnic Institute of Setúbal, Setúbal, Portugal
Ashish Ghosh
Indian Statistical Institute, Kolkata, India
Igor Kotenko
St. Petersburg Institute for Informatics and Automation of the Russian
Academy of Sciences, St. Petersburg, Russia
Takashi Washio
Osaka University, Osaka, Japan
Junsong Yuan
University at Buffalo, The State University of New York, Buffalo, USA
Lizhu Zhou
Tsinghua University, Beijing, China

More information about this series at http://www.springer.com/series/7899

Greg H. Parlier · Federico Liberatore ·
Marc Demange (Eds.)

Operations Research and Enterprise Systems

7th International Conference, ICORES 2018
Funchal, Madeira, Portugal, January 24–26, 2018
Revised Selected Papers

 Springer

Editors
Greg H. Parlier
INFORMS
Catonsville, MD, USA

Marc Demange
Royal Melbourne Institute of Technology
Melbourne, VIC, Australia

Federico Liberatore
UC3M-BS Institute of Financial Big Data
Universidad Carlos III de Madrid
Madrid, Spain

ISSN 1865-0929   ISSN 1865-0937 (electronic)
Communications in Computer and Information Science
ISBN 978-3-030-16034-0   ISBN 978-3-030-16035-7 (eBook)
https://doi.org/10.1007/978-3-030-16035-7

Library of Congress Control Number: 2019934744

© Springer Nature Switzerland AG 2019
This work is subject to copyright. All rights are reserved by the Publisher, whether the whole or part of the material is concerned, specifically the rights of translation, reprinting, reuse of illustrations, recitation, broadcasting, reproduction on microfilms or in any other physical way, and transmission or information storage and retrieval, electronic adaptation, computer software, or by similar or dissimilar methodology now known or hereafter developed.
The use of general descriptive names, registered names, trademarks, service marks, etc. in this publication does not imply, even in the absence of a specific statement, that such names are exempt from the relevant protective laws and regulations and therefore free for general use.
The publisher, the authors and the editors are safe to assume that the advice and information in this book are believed to be true and accurate at the date of publication. Neither the publisher nor the authors or the editors give a warranty, expressed or implied, with respect to the material contained herein or for any errors or omissions that may have been made. The publisher remains neutral with regard to jurisdictional claims in published maps and institutional affiliations.

This Springer imprint is published by the registered company Springer Nature Switzerland AG
The registered company address is: Gewerbestrasse 11, 6330 Cham, Switzerland

Preface

This book includes extended and revised versions of selected papers from the 7th International Conference on Operations Research and Enterprise Systems (ICORES 2018), held in Funchal, Madeira, Portugal, during January 24–26.

We received 59 paper submissions from 32 countries, of which 20% are included in this book. These papers were selected based on several criteria including reviews provided by Program Committee members, session chair assessments, and also program chair perspectives across all papers included in the technical program. The authors of these selected papers were then invited to submit revised and extended versions of their papers for formal publication.

The purpose of the annual ICORES conferences is to bring together researchers, engineers, and practitioners interested in both advances and applications in the field of operations research. Two simultaneous tracks are held, one covering domain-independent methodologies and technologies and the other practical work developed in specific application areas.

The papers selected for this book contribute to current research in operations research and a better understanding of complex enterprise systems. We commend each of the authors for their contributions, and gratefully thank our many reviewers who ensured the high quality of this publication.

January 2018

Greg H. Parlier
Federico Liberatore
Marc Demange

Organization

Conference Chair

Marc Demange   RMIT University, Australia

Program Co-chairs

Greg H. Parlier   NCSU, USA
Federico Liberatore   Universidad Complutense de Madrid, Spain

Program Committee

El-Houssaine Aghezzaf	Ghent University, Belgium
Maria Teresa Almeida	ISEG, Universidade de Lisboa, Portugal
Cláudio Alves	Universidade do Minho, Portugal
Lionel Amodeo	University of Technology of Troyes, France
Necati Aras	Bogazici University, Turkey
Eduardo Barbosa	Brazilian National Institute for Space Research (Inpe), Brazil
Patrizia Beraldi	University of Calabria, Italy
David Bergman	University of Connecticut, USA
Giancarlo Bigi	University of Pisa, Italy
Peter Vanden Bosch	Marymount University, USA
Renato Bruni	University of Rome La Sapienza, Italy
Ahmed Bufardi	NA, Switzerland
Sujin Bureerat	KhonKaen University, Thailand
Alfonso Mateos Caballero	Universidad Politécnica de Madrid, Spain
Olivier Caelen	Atos Worldline, Belgium
Massimiliano Caramia	University of Rome Tor Vergata, Italy
Mirko Cesarini	University of Milano-Bicocca, Italy
Bo Chen	University of Warwick, UK
Andre Augusto Cire	University of Toronto, Canada
Roberto Cordone	University of Milan, Italy
Florbela Maria Cruz Domingues Correia	Instituto Politécnico de Viana do Castelo, Portugal
Heliodoro Daniel Cruz-Suárez	Universidad Juárez Autónoma de Tabasco, Mexico
Patrizia Daniele	University of Catania, Italy
Mirela Danubianu	Stefan cel Mare University of Suceava, Romania
Andrea D'Ariano	Università degli Studi Roma Tre, Italy
Marc Demange	RMIT University, Australia
Clarisse Dhaenens	CRIStAL, France

Khaled Elbassioni	Masdar Institute, Khalifa University of Science and Technology, UAE
Ali Emrouznejad	Aston University, UK
Nesim Erkip	Bilkent University, Turkey
Laureano F. Escudero	Universidad Rey Juan Carlos, Spain
Mathilde Excoffier	LRI, Université Paris-Sud XI (Orsay), France
Luis Miguel D. F. Ferreira	Universidade de Coimbra, Portugal
Paola Festa	University of Naples, Italy
Ingrid Fischer	Universität Konstanz, Germany
Muhammad Marwan Muhammad Fuad	Technical University of Denmark, Denmark
Claudio Gentile	CNR, Italy
Ronald Giachetti	Naval Postgraduate School, USA
Ilias Gialampoukidis	Information Technologies Institute, Greece
Stefano Giordani	University of Rome Tor Vergata, Italy
Alessandro Giuliani	University of Cagliari, Italy
Giorgio Gnecco	IMT, School for Advanced Studies Lucca, Italy
Marc Goerigk	Lancaster University, UK
Boris Goldengorin	Ohio University, USA
Dries Goossens	Ghent University, Belgium
Stefano Gualandi	University of Pavia, Italy
Christelle Guéret	University of Angers, France
Francesca Guerriero	University of Calabria, Italy
Gregory Z. Gutin	Royal Holloway University of London, UK
Jin-Kao Hao	University of Angers, France
Kenji Hatano	Doshisha University, Japan
Hanno Hildmann	Universidad Carlos III de Madrid, Spain
Chenyi Hu	The University of Central Arkansas, USA
Johann Hurink	University of Twente, The Netherlands
Josef Jablonsky	University of Economics, Czech Republic
Antonio Jiménez-Martín	Universidad Politécnica de Madrid, Spain
Itir Karaesmen	American University, USA
Daniel Karapetyan	University of Essex, UK
Michael Katehakis	Rutgers University, USA
George Katsirelos	INRA, France
W. David Kelton	University of Cincinnati, USA
Ahmed Kheiri	Lancaster University, UK
Philip Kilby	NICTA and the Australian National University, Australia
Jesuk Ko	Gwangju University, Korea, Republic of
Michal Koháni	University of Zilina, Slovak Republic
Sotiria Lampoudi	Droneseed, USA
Dario Landa-Silva	University of Nottingham, UK
Richard C. Larson	Massachusetts Institute of Technology, USA
Pierre L'Ecuyer	Universite de Montreal, Canada
Janny Leung	The Chinese University of Hong Kong, SAR China

Benjamin Lev	Drexel University, USA
Dimitrios Liparas	High Performance Computing Center Stuttgart (HLRS), Germany
Pierre Lopez	LAAS-CNRS, Université de Toulouse, France
Qiang Ma	Kyoto University, Japan
Prabhat K. Mahanti	University of New Brunswick, Canada
Viliam Makis	University of Toronto, Canada
Arnaud Malapert	Université Côte d'Azur, CNRS, I3S, France
Emanuele Manni	University of Salento, Italy
Fabrizio Marinelli	Università Politecnica delle Marche, Italy
Concepción Maroto	Universidad Politécnica de Valencia, Spain
Pedro Coimbra Martins	Polytechnic Institute of Coimbra, Portugal
Carlo Meloni	Politecnico di Bari, Italy
Marta Mesquita	Universidade de Lisboa, Portugal
Rym MHallah	Kuwait University, Kuwait
Jairo R. Montoya-Torres	Universidad de los Andes, Colombia
Young Moon	Syracuse University, USA
Gaia Nicosia	Università degli Studi Roma Tre, Italy
Inneke Van Nieuwenhuyse	KU Leuven, Belgium
José Oliveira	Universidade do Minho, Portugal
Pedro Nuno Ferreira Pinto Oliveira	Universidade do Porto, Portugal
Mohammad Oskoorouchi	California State University, San Marcos, USA
Selin Özpeynirci	Izmir University of Economics, Turkey
Massimo Paolucci	University of Genoa, Italy
Mauro Passacantando	University of Pisa, Italy
Gabrielle Peko	The University of Auckland, New Zealand
Ulrich Pferschy	University of Graz, Austria
Caroline Prodhon	Charles Delaunay Institute, France
Luca Quadrifoglio	Texas A&M University, USA
Günther Raidl	Vienna University of Technology, Austria
Celso Ribeiro	Universidade Federal Fluminense, Brazil
Michela Robba	University of Genoa, Italy
Helena Sofia Rodrigues	Instituto Politécnico de Viana do Castelo, Portugal
Andre Rossi	Université d'Angers, France
Lukas Ruf	Consecom AG, Switzerland
Stefan Ruzika	University of Kaiserslautern, Germany
Mohamed Saleh	Cairo University, Egypt
Cem Saydam	University of North Carolina Charlotte, USA
Abdelkader Sbihi	EM Normandie, France
Andrea Scozzari	Università degli Studi Niccolo' Cusano, Italy
Laura Scrimali	University of Catania, Italy
René Séguin	Defence Research Development Canada, Canada
Patrick Siarry	University of Paris 12 (LiSSi), France
Fabio Stella	University of Milano-Bicocca, Italy
Thomas Stützle	Université Libre de Bruxelles, Belgium

Wai Yuen Szeto	The University of Hong Kong, SAR China
Vadim Timkovski	South University, USA
Norbert Trautmann	University of Bern, Switzerland
Chefi Triki	University of Salento, Italy
Michael Tschuggnall	University of Innsbruck, Austria
Maria Vlasiou	Eindhoven University of Technology, The Netherlands
Cameron Walker	University of Auckland, New Zealand
Santoso Wibowo	CQUniversity, Australia
Gerhard Woeginger	Eindhoven University of Technology, The Netherlands
Neil Yorke-Smith	TU Delft, The Netherlands
Xufeng Zhao	Nanjing University of Aeronautics and Astronautics, China
Yiqiang Zhao	Carleton University, Canada
Jan Zizka	Mendel University in Brno, Czech Republic
Paola Zuddas	University of Cagliari, Italy

Additional Reviewers

| Xuan Thuy Pham | LiSSi Laboratory, France |
| Slawomir Wesolkowski | Defence Research and Development Canada, Canada |

Invited Speakers

Aharon Ben-Tal	Technion, Israel Institute of Technology, Israel
Aaron Burciaga	VP Data Science and Machine Learning, Booz Allen Hamilton, USA
Abdelkader Sbihi	Normandy University, France

Contents

Applications

Methodologies and Technologies

A Stochastic Production Frontier Analysis of the Brazilian Agriculture in the Presence of an Endogenous Covariate

Geraldo da Silva e Souza$^{(\boxtimes)}$ and Eliane Gonçalves Gomes

Secretaria de Inteligência e Relações Estratégicas – Embrapa,
Brasília, DF 70770-901, Brazil
{geraldo.souza,eliane.gomes}@embrapa.br

Abstract. Production frontier analysis aims at the identification of best production practices and the importance of external factors, endogenous or not, that affect the production function and the technical efficiency component. In particular, in the context of the Brazilian agriculture, it is desirable for policy makers to identify the effect on production of variables related to market imperfections. Market imperfections occur when farmers are subjected to different market conditions depending on their income. In general, large scale farmers access lower input prices and may sell their production at lower prices, thereby making competition harder for small farmers. Market imperfections are typically associated with infrastructure, environment control requirements and the presence of technical assistance. In this article, at county level, and using agricultural census data, we estimate the elasticities of these variables on production by maximum likelihood methods. Technological inputs dominate the production response, followed by labor and land. Environment control has a positive net effect on production, as well as technical assistance. The indicator of infrastructure affects positively technical efficiency. There is no evidence of technical assistance endogeneity.

Keywords: Stochastic frontier · Endogeneity · Agriculture

1 Introduction

As pointed out in other sources [1–3], Brazilian agriculture is highly concentrated. Only five hundred thousand farmers, 11.4% of the total, produced 87% of the total production value in 2006 (last agricultural census). These data motivate studies that identify factors of importance for public policies leading to productive inclusion in agriculture in Brazil. Indeed, the major (state) agricultural research company in Brazil defines "productive insertion and poverty reduction" as one of the impact axes in its strategic planning map. Access to technology is the main cause of production concentration and, very likely, of poverty in the fields. We see, in this context, that the agricultural sector demands proper public policies in order to improve access to technology and to increase productive insertion and reduce rural poverty.

As emphasized in Souza and Gomes [3], market imperfections are the main cause of inhibition of the access of farmers to technology and, therefore, to productive

© Springer Nature Switzerland AG 2019
G. H. Parlier et al. (Eds.): ICORES 2018, CCIS 966, pp. 3–14, 2019.
https://doi.org/10.1007/978-3-030-16035-7_1

inclusion. Market imperfections are the result of asymmetry in credit for production, infrastructure, information availability, rural extension and technical assistance, among others [4].

The lack of physical infrastructure and education make it difficult for the rural extension to fulfill its role and, therefore, gain proper access to technology. Another point to be emphasized is related to the imperfection of the production markets. Souza et al. [5] highlight that small farmers sell their products at lower values and buy inputs at higher prices. The large scale producers are able to negotiate better input and output prices and the existence of these different prices characterizes a market imperfection. Unfavorable negotiation may lead to higher prices for the adoption of better technologies and thus lead to difficulties in achieving higher economic efficiency.

We contribute to this literature modeling production value as a function of several aggregates, reflecting, on a municipal level, the input usage, environment control, technical assistance and the effect of market imperfection variables on the technical efficiency of production. The modeling process postulates a Cobb-Douglas representation in a typical stochastic frontier approach and is carried out under the assumption of endogeneity of technical assistance. The models we used follow the basic lines of Karakaplan and Kutlu [6] and Karakaplan [7]. We extend Karakaplan's approach to the truncated normal and the exponential distributions. Alternatives to the approach are also suggested, considering non-linear models with the Murphy and Topel [8] variance correction. In this context we allow the use of fractional regressions [9, 10]. Our results extend Souza and Gomes' [3] findings.

2 Data

The data sources for this article are the Brazilian agricultural census of 2006, the Brazilian demographic census of 2010, and municipal databases on education and health.

We follow the approach of Souza et al. [2, 5] in the definition of production and contextual variables.

Production (inputs and output) is defined using monetary values. The source is the agricultural census of 2006 [11]. The output variable is the value of production and the inputs are expenses on labor (*labor*), land (*land*) and technological inputs (*techinputs*), which includes machinery, improvements in the farm, equipment rental, value of permanent crops, value of animals, value of forests in the establishment, value of seeds, value of salt and fodder, value of medication, fertilizers, manure, pesticides, expenses with fuel, electricity, storage, services provided, raw materials, incubation of eggs and other expenses. Value of permanent crops, forests, machinery, improvements on the farm, animals and equipment rental were depreciated at a rate of 6% a year (machines – 15 years, planted forests – 20 years, permanent cultures – 15 years, improvements – 50 years, animals – 5 years). Farm data from the agricultural census were aggregated to form totals for each county. A total of 4,965 counties (almost 90% of the total) provided valid data for our analysis.

The contextual variables we chose are a performance municipal index of social development (*social*), an index of demographic development (*demog*), the proportion of farmers who received technical assistance (*techassist*), the proportion of non-degraded areas (*ndareas*) and the proportion of forested areas (*forest*). The last two are proxies for environment control. Market imperfections are mainly associated with the social index.

The demographic index captures the population dynamics that tend to follow rural development. The variables considered in this dimension of development are the migration index (rural to urban areas), average number of farm dwellers, aging rate (total municipal rural population over rural population over 60), dependency rate (ratio of the rural population with age in the bracket 15–59 over the rural population with age in the bracket of 0–14 plus over 60), ratio of urban to rural population in the municipality. The source is the demographic census of 2010 [12] in general, and the 2000 and 2010 census for the migration index. The demographic score was computed using the ranks of these measurements, weighted by the relative multiple correlation coefficient.

The index of social development reflects the level of well-being, favored by factors such as the availability of water and electric energy in the rural residences, and level of education, health and poverty in the rural households. It was computed as a weighted average of normalized ranks of the following variables: education (illiteracy rate), poverty index, average gross per capita income of rural households, proportion of farms with access to electricity and water, index of basic education, index of performance of the public health system and vulnerability of children up to five years old. These indicators were obtained from the Brazilian demographic census 2010 [12], from the Brazilian agricultural census 2006 [11], and from the databases of the National Institute of Research and Educational Studies (INEP), referring to education in 2009 [13], and of the Ministry of Health 2011 data [14]. The social score was computed using the ranks of these measurements, weighted by the relative multiple correlation coefficient.

3 Methodology

Our approach to assess production and efficiency of production follows along the lines of Karakaplan and Kutlu [6] and Karakaplan [7]. The structural model for our application is defined by (1) for municipality i, where *techassist* is assumed endogenous and y_i is the log of gross income.

$$y_i = b_0 + b_1 \log(labor_i) + b_2 \log(land_i) + b_3 \log(techinputs_i) + b_4(forext_i)$$
$$+ b_5(ndareas_i) + b_6(techassist_i) + v_i - u_i$$
$$v_i, u_i \text{ independent} \tag{1}$$
$$v_i \sim N(0, \sigma^2)$$

The u_i are non-negative inefficiency components and the v_i are a random sample of an idiosyncratic error component. We assume three possible distributions for the inefficiency component: half-normal, exponential and truncated normal.

For the half-normal we have $u_i \sim N^+\left(0, \sigma_{u_i}^2\right)$ and $\sigma_{u_i}^2 =$ $\exp\left(\begin{array}{l} b_7 + b_8 \log(labor_i) + b_9 \log(land_i) + b_{10} \log(techinputs_i) + b_{11}(forext_i) + \\ b_{512}(ndareas_i) + b_{13}(social_i) + b_{14}(demog_i) \end{array}\right)$. For the exponential $u_i \sim \exp(\zeta_i), \zeta > 0$, we assume the variance $\sigma_{u_i}^2 = \zeta^{-2}$ with the same representation as the half-normal. Finally, for the truncated normal $u_i \sim N^+\left(\mu_i, \sigma_u^2\right)$ and $\mu_i = b_7 + b_8 \log(labor_i) + b_9 \log(land_i) + b_{10} \log(techinputs_i) + b_{11}(forext_i) + b_{512}(ndareas_i) + b_{13}(social_i) + b_{14}(demog_i)$.

Endogeneity in Karakaplan and Kutlu [6] and Karakaplan [7] means correlation of a variable with v_i. This assumption invalidates the classic stochastic frontier analysis. A classic approach for handling this issue is to use two stage least squares or the general method of moments (GMM), as suggested in Amsler et al. [15]. On the other hand, Karakaplan and Kutlu [6] and Karakaplan [7] suggest the use of instrumental variables in a context of maximum likelihood estimation, resembling classical frontier analysis. In our application, we follow this approach and the instruments considered for *techassist* are the exogenous variables plus demographic and social indicators. The instrumental variable regression is assumed to be linear but the idea can be easily generalized to non-linear specifications $techassist_i = f(z_i, \delta) + \varepsilon_i$. In this formulation, z_i is a vector of instrumental variables and $\varepsilon' = (\varepsilon_1, \ldots, \varepsilon_n)$ has mean zero and variance matrix $\sigma_\varepsilon^2 I$. Heteroskedastic formulations are possible assuming a general variance matrix Ω. This formulation also allows for the Bernoulli specification described in Papke and Wooldridge [9], which is particularly convenient if one is dealing with fractions. In this instance, the model can be estimated assuming $f(.)$ to be a distribution function.

Karakaplan [7] in its 'sfkk' module in the Stata software makes use of the half-normal distribution and the linear instrumental variable regression.

Let ρ be the correlation between ε_i and v_i. Endogeneity means $\rho \neq 0$. We assume the bivariate normal distribution as in (2).

$$\begin{bmatrix} \tilde{\varepsilon}_i \\ v_i \end{bmatrix} = \begin{bmatrix} \varepsilon_i/\sigma_\varepsilon \\ v_i \end{bmatrix} \sim N\left(\begin{bmatrix} 0 \\ 0 \end{bmatrix}, \begin{bmatrix} 1 & \rho\sigma \\ \rho\sigma & \sigma^2 \end{bmatrix}\right) \tag{2}$$

Using a Cholesky decomposition we may write (3) and, therefore, we have (4).

$$\begin{bmatrix} \tilde{\varepsilon}_i \\ v_i \end{bmatrix} = \begin{bmatrix} 1 & 0 \\ \rho\sigma & \sigma\sqrt{1-\rho^2} \end{bmatrix} \begin{bmatrix} \tilde{\varepsilon}_i \\ \tilde{w}_i \end{bmatrix} \tag{3}$$

$$\begin{aligned} y_i &= b_0 + b_1 \log(labor_i) + b_2 \log(land_i) + b_3 \log(techinputs_i) + b_4(forext_i) \\ &\quad + b_5(ndareas_i) + b_6(techassist_i) + \eta\tilde{\varepsilon}_i + w_i - u_i \\ w_i &= \sigma\sqrt{1-\rho^2}\tilde{w}_i \\ \mu &= \rho\sigma \end{aligned} \tag{4}$$

Therefore, when the residual variance is constant, the component $\eta\tilde{\varepsilon}_i$ is the correction term for bias. The test of $\eta = 0$ is an endogeneity test. The model is estimated by maximum likelihood.

For the half-normal distribution, the likelihood function is given by (5).

$$\log L(\theta) = \sum_{i=1}^{n} \left\{ \frac{\ln(2/\pi) - \ln(\sigma_{Si}^2) - (e_i/\sigma_{Si}^2)}{2} + \ln \Phi\left(\frac{\lambda_i e_i}{\sigma_{Si}}\right) \right\}$$
$$+ \sum_{i=1}^{n} \left\{ \frac{\ln 2\pi - \ln \sigma_\varepsilon - \sum_{i=1}^{n}\left(\varepsilon_i^2/\sigma_\varepsilon^2\right)}{2} \right\} \tag{5}$$

Here $\lambda_i = \sigma_{ui}/\sigma$ and $\sigma_{Si}^2 = \sigma_{ui}^2 + \sigma^2$. Notice that e_i is defined by (6).

$$e_i = y_i - \left(\begin{array}{l} b_0 + b_1 \log(labor_i) + b_2 \log(land_i) + b_3 \log(techinputs_i) + b_4(forext_i) \\ + b_5(ndareas_i) + b_6(techassist_i) + \eta\tilde{\varepsilon}_i \end{array} \right) \tag{6}$$

For the exponential model, the likelihood function becomes (7) and for the truncated normal it is defined as in (8), where $\gamma = \sigma_u^2/\sigma_S^2$, $\sigma_S^2 = \sigma_u^2 + \sigma^2$.

$$\log L(\theta) = \sum_{i=1}^{n} \left\{ -\ln(\sigma_u) + \frac{\sigma^2}{2\sigma_u^2} + \ln \Phi\left(\frac{-e_i - \sigma^2/\sigma_u}{\sigma}\right) + \frac{e_i}{\sigma_u} \right\}$$
$$+ \sum_{i=1}^{n} \left\{ \frac{\ln 2\pi - \ln \sigma_\varepsilon - \sum_{i=1}^{n}\left(\varepsilon_i^2/\sigma_\varepsilon^2\right)}{2} \right\} \tag{7}$$

$$\log L(\theta) = \sum_{i=1}^{n} \left\{ \begin{array}{l} -\dfrac{\ln(2\pi)}{2} - \ln \sigma_S - \ln \Phi\left(\dfrac{\mu_i}{\sigma_S\sqrt{\gamma}}\right) + \ln \Phi\left(\dfrac{(1-\gamma)\mu_i - \lambda e_i}{\sigma_S\sqrt{\gamma(1-\gamma)}}\right) \\[2mm] -\dfrac{1}{2}\left(\dfrac{e_i + \mu_i}{\sigma_S}\right)^2 \end{array} \right\}$$
$$+ \sum_{i=1}^{n} \left\{ \frac{\ln 2\pi - \ln \sigma_\varepsilon - \sum_{i=1}^{n}\left(\varepsilon_i^2/\sigma_\varepsilon^2\right)}{2} \right\} \tag{8}$$

Karakaplan and Kutlu [6] suggest an alternative to estimation easier to implement, which can be extended to accommodate fractional regression models in the instrumental regression. The idea is to perform the estimation in two steps. Firstly, one fits the instrumental variable regression and computes residuals $\hat{\varepsilon}_i = techassist - f\left(z_i, \hat{\delta}\right)$ and then runs the standard stochastic frontier model (9).

$$y_i = b_0 + b_1 \log(labor_i) + b_2 \log(land_i) + b_3 \log(techinputs_i) + b_4(forext_i)$$
$$+ b_5(ndareas_i) + b_6(techassist_i) + \eta\hat{\varepsilon}_i + w_i - u_i \tag{9}$$

The process will not produce the same results as the full maximum likelihood estimation. Greene [4] names it limited information maximum likelihood. The variance matrix of the estimator requires the Murphy and Topel [8] correction. Let $\hat{\delta}$ be the maximum likelihood estimate obtained from the instrumental variable regression with variance matrix $\hat{V}_1$. The likelihood function is $\ln(techassist, z, \delta)$. Let $\hat{\theta}$ be the maximum likelihood of the resulting frontier model obtained when $\delta = \hat{\delta}$. The variance matrix is $\hat{V}_2$ and the likelihood function is $\ln f_2\left(y, x, \hat{\delta}, \theta\right)$, where the vector x includes inputs, technical assistance, non-degraded areas, forests, and the residual from the instrumental variable regression. Following Greene [16], we may define the matrices (10) and (11).

$$\hat{C} = \frac{1}{n} \sum_{i=1}^{n} \left(\frac{\partial \ln f_{2i}}{\partial \hat{\theta}}\right)\left(\frac{\partial \ln f_{2i}}{\partial \hat{\delta}'}\right) \tag{10}$$

$$\hat{R} = \frac{1}{n} \sum_{i=1}^{n} \left(\frac{\partial \ln f_{2i}}{\partial \hat{\theta}}\right)\left(\frac{\partial \ln f_{1i}}{\partial \hat{\delta}'}\right) \tag{11}$$

The estimated variance matrix of the limited information maximum likelihood estimator is defined as in (12).

$$\hat{V} = \frac{1}{n}\left(\hat{V}_2 + \hat{V}_2\left(\hat{C}\hat{V}_1\hat{C}' - \hat{R}\hat{V}_1\hat{C}' - \hat{C}\hat{V}_1\hat{R}'\right)\hat{V}_2\right) \tag{12}$$

In our exercise we used both methods, that is, full likelihood estimation as well as the two-step procedure. Regression in the first step used both the fractional approach of Papke and Wooldridge [9] and linear regression.

4   Statistical Results

Following the standard literature in stochastic frontier analysis we fitted 11 models to the data described in Sect. 2, using the approaches of Sect. 3. The models considered are: Case 1 – The full information maximum likelihood approach under the half-normal and exponential inefficiency distributions, and the correspondent limited information maximum likelihood for the best model under linear and fractional instrumental variables regressions. The only inefficiency effect considered is the social indicator; Case 2 – The limited information maximum likelihood assuming both instrumental variables' regression assumptions, including as efficiency effects all independent factors for the half-normal and truncated normal. Tables 1 and 2 show the goodness of fit measures considered for model choice.

Table 1. Fit statistics: Case 1 – Social indicator is the only inefficiency effect.

Model	Inefficiency distribution	Likelihood
Full information ML	Half-normal	−3041.3
Full information ML	Truncated normal	Do not converge
Full information ML	Exponential	−3042.8
Limited information ML – linear	Half-normal	−4663.3
Limited information ML – fractional	Half-normal	−4661.8

Table 2. Fit statistics: Case 2 – All exogenous variables are inefficiency effects.

Model	Inefficiency distribution	Likelihood
Limited information ML – fractional	Half-normal	−4603.4
Limited information ML – fractional	Truncated normal	−4607.4
Limited information ML – fractional	Exponential	Do not converge
Limited information ML – linear	Half-normal	−4603.4
Limited information ML – linear	Truncated normal	−4612.0
Limited information ML – linear	Exponential	Do not converge

We experienced convergence problems with some of the assumptions for the inefficiency distribution, depending on the assumption itself and on the number of variables included in the efficiency effect. The full information maximum likelihood with all exogenous variable included in the effect did not converge, inhibiting the application of the standard likelihood approach to test nested hypothesis. We see from Tables 1 and 2 that the best fit is the full information estimator under the half-normal distribution, reducing the set of inefficiency factor effects to the social indicator. The models fitted in two stages using the linear and the non-linear binomial Papke and Woodridge [12] assumptions indicate similar results, with a slight superiority for the fractional regression. Correlations between actual and estimated values for the instrumental regressions are, respectively, 80.1% and 80.4%. Programming was carried out using Stata 14 and SAS 9.2 software.

Table 3 shows statistical estimation for full information half-normal model including a social effect for the inefficiency component. Table 4 shows the fractional regression for technical assistance. Table 5 shows the limited information maximum likelihood with the Murphy-Topel variance correction [8], under the binomial specification for the instrumental variable regression.

In the context of the full information maximum likelihood estimation correlation between actual and predicted values of the frontier model, including efficiency effects, is 88.6%. The component technical assistance affects significantly and positively the response variable (log income). There is no evidence of endogeneity (p-value = 0.1858).

Table 6 summarizes the relative importance of production factors, including returns to scale. We see that technology dominates, followed by labor and land. The technology shows decreasing returns to scale. These results fairly agree with Souza et al. [5].

Table 3. Full information maximum likelihood estimation. Half-normal stochastic frontier under endogeneity of technical assistance. Stata output.

	Coefficient	Std error	z	P > |z|	[95% confidence interval]	
Frontier						
labor	0.231137	0.011531	20.04	0.000	0.208536	0.253738
land	0.09003	0.013968	6.45	0.000	0.062653	0.117406
techinputs	0.45581	0.021104	21.6	0.000	0.414446	0.497173
forest	−0.12398	0.032878	−3.77	0.000	−0.18842	−0.05954
ndareas	0.250139	0.036281	6.89	0.000	0.17903	0.321249
techassist	0.567809	0.140459	4.04	0.000	0.292514	0.843105
constant	2.736811	0.104023	26.31	0.000	2.53293	2.940691
Instrumental regression						
labor	−0.02131	0.003139	−6.79	0.000	−0.02747	−0.01516
land	0.007906	0.003929	2.01	0.044	0.000207	0.015606
techinputs	0.077737	0.004742	16.39	0.000	0.068443	0.087031
forest	0.020425	0.009285	2.2	0.028	0.002227	0.038624
ndareas	0.086496	0.008944	9.67	0.000	0.068967	0.104026
social	0.659066	0.015642	42.14	0.000	0.628409	0.689723
demog	−0.12634	0.028992	−4.36	0.000	−0.18316	−0.06952
constant	−0.44813	0.023053	−19.44	0.000	−0.49331	−0.40294
η						
constant	−0.1976	0.149364	−1.32	0.186	−0.49035	0.095144
$\ln \sigma_u^2$						
social	−2.17789	0.737983	−2.95	0.003	−3.62432	−0.73147
constant	−2.47837	0.762352	−3.25	0.001	−3.97255	−0.98419
$\ln \sigma^2$						
constant	−0.9899	0.027306	−36.25	0.000	−1.04341	−0.93638

η Endogeneity Test: Ho: Correction for endogeneity is not necessary; Ha: There is endogeneity in the model and correction is needed.

(1) $[\eta\_techassist]\_constant = 0$
$\chi^2(1) = 1.75$
Prob $> \chi^2 = 0.1858$
Result: Cannot reject Ho at 10% level.

Technical assistance, non-degraded areas and the proportion of forested areas are all statistically significant (Table 4). The former act favoring production and the latter have a negative effect.

Table 7 shows 5-number summaries for technical efficiency. Figure 1 shows box plots for the normalized ranks of the efficiency measurements. Efficiency differs significantly by regional classification. There is a clear domination of South, Southeast, and Center-West.

Table 4. Instrumental variable fractional regression. SAS output.

Parameter	Estimate	Error	DF	t Value	Pr > |t|	[95% confidence interval]	
Labor	−0.0579	0.0244	4965	−2.37	0.0178	−0.1058	−0.0100
land	0.0160	0.0306	4965	0.52	0.6008	−0.0440	0.0761
techinputs	0.2375	0.0374	4965	6.35	<.0001	0.1642	0.3108
forest	0.0555	0.0718	4965	0.77	0.4394	−0.0853	0.1964
ndareas	0.2609	0.0692	4965	3.77	0.0002	0.1253	0.3965
social	1.8260	0.1217	4965	15.01	<0.0001	1.5874	2.0645
demog	−0.4441	0.2226	4965	−2.00	0.0461	−0.8805	−0.0078
constant	−2.7670	0.1840	4965	−15.04	<0.0001	−3.1277	−2.4063

Table 5. Limited Information maximum likelihood estimation. Half-normal stochastic frontier under endogeneity of technical assistance and fractional regression. SAS output.

Parameter	Estimate	Std error frontier model	Std error Murphy-Topel	Murphy-Topel 95% confidence interval		P > |z|
Frontier						
labor	0.2320	0.01139	0.01023	0.21196	0.25204	0.00000
land	0.0880	0.01394	0.01108	0.06623	0.10967	0.00000v
techinputs	0.4522	0.02027	0.01829	0.41635	0.48805	0.00000
techassist	0.6050	0.12580	0.16852	0.27470	0.93530	0.00033
forest	−0.1230	0.03280	0.03011	−0.18202	−0.06398	0.00004
ndareas	0.2424	0.03558	0.03517	0.17346	0.31134	0.00000
residual	−0.2529	0.13590	0.17634	−0.59853	0.09273	0.15153
constant	2.7732	0.10250	0.12101	2.53602	3.01038	0.00000
$\ln \sigma_u^2$						
social	−2.0794	0.6869	2.47472	−6.92985	2.77105	0.40076
constant	−2.4200	0.62090	0.83613	−4.05881	−0.78119	0.0038
$\ln \sigma^2$						
constant	0.3702	0.00955	0.01528	0.34026	0.40014	0.00000

The social indicator positively affects technical efficiency, as reported in Table 4. Regions that are to benefit the most from improvements in the social indicators are the North and Northeast. The instrumental variable regression indicates a strong dependence of technical assistance on the environment, demographics and the social conditions. The increased population dynamics makes the presence of technical assistance unnecessary, implying, therefore, a negative effect of the demographic index. The other indices are positively related to technical assistance.

Table 6. Relative elasticities and returns to scale.

Production factor	Relative elasticity	Standard error
Labor	0.297	0.016
Land	0.116	0.018
Technology	0.587	0.022
Returns to Scale	0.777	0.014

Table 7. Normalized rank of technical efficiency – 5-number summary.

Region	Minimum	Q1	Median	Q3	Maximum
North	0.000	0.192	0.313	0.425	0.957
Northeast	0.000	0.098	0.206	0.344	0.999
Southeast	0.004	0.502	0.695	0.854	1.000
South	0.055	0.618	0.744	0.866	1.000
Center-west	0.012	0.405	0.526	0.543	0.990

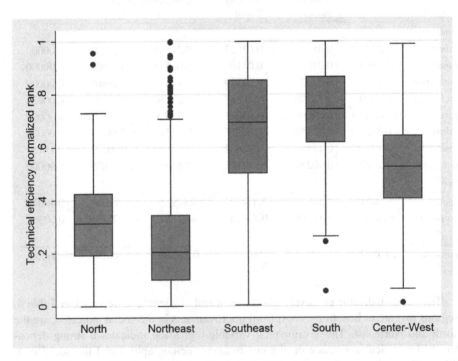

Fig. 1. Box plots of technical efficiency by region.

The limited information maximum likelihood estimation agrees, in general, with the full information maximum likelihood results. There is no evidence of technical assistance endogeneity. See Table 5. The main difference regards the standard error of the estimated coefficient of the social indicator in the inefficiency variance (Table 5). The Murphy-Topel correction inflates the variance, forcing non-significance. However, the coefficient values are similar. The fractional instrumental variables regression indicates positive relation to the social indicator and to non-degraded areas (Table 4). The demographic index is negatively related to the response and the proportion of forested areas is not significant.

Limited information estimation, including all instrumental variables as technical efficiency effects, is clearly inferior to the full information model estimated, including only the social indicator (Tables 1 and 2). The interesting feature of these models is the similarity of the results obtained with the linear and non-linear instrumental regression, suggesting robustness of the linear instrumental regression.

5 Concluding Remarks

We fitted a stochastic frontier under endogeneity to municipal data using the Brazilian agricultural census of 2006 – the last one available. The objective of this study, besides assessing input elasticities, was to investigate effects of market imperfection variables on production. Market imperfections come from different realities in production experienced by small and large farmers. They relate to infrastructure, level of education and access to credit, implying in different input and output prices for small and large farmers. The presence of market imperfection makes it harder for rural extension and technical assistance to promote productive inclusion.

For public policy decision-making, the identification of production function components elasticities is of importance to guide rural governmental assistance. This is critical for reducing poverty in the fields and for increasing production. We conclude that technology is the main input factor for increasing income in rural Brazil. The social indicator is the key variable to reducing inefficiency. The indicator is relatively too low for the Northern and Northeastern regions. Values are less than half of the corresponding values of other regions. Public policies should be oriented to improve this indicator particularly in these regions.

Technical assistance is an important part of rural extension and has a direct positive effect on income. Improvement of the social indicator will tend to facilitate the access of technical assistance creating, in this way, a synergic positive effect on income.

Environment in our study was measured in two ways: non-degraded areas and the proportion of forested areas. Keeping non-degraded areas relates to technology and has a positive impact on production. On the other hand, keeping a relative large area of uncultivated land in the farm will have a negative effect on income. Extension and technical assistance may be the key factor to extract value from forests and properly preserve these areas.

Finally, we emphasize the fact that the use of limited information maximum likelihood estimation indeed eases convergence in the stochastic frontier models. The linear instrumental regression seems to be robust, but it produces inferior fits when compared with fractional regressions. The Murphy-Topel variance matrix correction may change the significance of important variables relative to the full information maximum likelihood estimation.

References

1. Alves, E., Souza, G.S., Rocha, D.P.: Desigualdade nos campos sob a ótica do censo agropecuário 2006. Revista de Política Agrícola 22, 67–75 (2013)
2. Souza, G.S., Gomes, E.G., Alves, E.R.A., Magalhães, E., Rocha, D.P.: Um modelo de produção para a agricultura brasileira e a importância da pesquisa da Embrapa. In: Alves, E. R.A., Souza, G.S., Gomes, E.G. (eds.) Contribuição da Embrapa para o desenvolvimento da agricultura no Brasil, pp. 49–86. Embrapa, Brasília (2013)
3. Souza, G.S., Gomes, E.G.: The effect of marketing imperfection variables on production in the context of Brazilian agriculture. In: Proceedings of the 7th International Conference on Operations Research and Enterprise Systems (ICORES 2018), pp. 15–20, Scitepress, Setúbal (2018)
4. Alves, E., Souza, G.S.: Pequenos estabelecimentos em termos de área também enriquecem? Pedras e tropeços. Revista de Política Agrícola 24, 7–21 (2015)
5. Souza, G.S., Gomes, E.G., Alves, E.R.A.: Conditional FDH efficiency to assess performance factors for Brazilian agriculture. Pesquisa Operacional 37, 93–106 (2017)
6. Karakaplan, M.U., Kutlu, L.: Handling endogeneity in stochastic frontier analysis (2013). http://www.mukarakaplan.com/Karakaplan%20-%20EndoSFA.pdf. Accessed 10 Mar 2017
7. Karakaplan, M.U.: Fitting endogenous stochastic frontier models in Stata. Stata J. 17(1), 39–55 (2017)
8. Murphy, K.M., Topel, R.H.: Estimation and inference in two step econometric models. J. Bus. Econ. Stat. 3, 370–379 (1985)
9. Papke, L.E., Wooldridge, J.M.: Econometric methods goes fractional response variables with an application to 401(k) plan participation rates. J. Appl. Econ. 11(6), 619–632 (1996)
10. Ramalho, E.A., Ramalho, J.J.S., Henriques, P.D.: Fractional regression models for second stage DEA efficiency analyses. J. Prod. Anal. 34, 239–255 (2010)
11. IBGE Homepage. Censo Agropecuário (2006). http://www.ibge.gov.br/home/estatistica/economia/agropecuaria/censoagro/. Accessed 24 Jan 2012
12. IBGE Homepage. Censo Demográfico (2010). http://censo2010.ibge.gov.br/. Accessed 24 Jan 2012
13. INEP Homepage. Nota Técnica do Índice de Desenvolvimento da Educação Básica (2012) http://ideb.inep.gov.br/resultado/. Accessed 24 Jan 2012
14. Ministério da Saúde Homepage. IDSUS – Índice de Desempenho do SUS (2011). http://portal.saude.gov.br/. Accessed 02 Mar 2012
15. Amsler, C., Prokhorov, A., Schmidt, P.: Endogeneity in stochastic frontier models. J. Econometrics 190, 280–288 (2016)
16. Greene, W.H.: Econometric Analysis, 6th edn. Prentice Hall, Englewood Cliffs (2008)

Dynamic Pricing Competition with Unobservable Inventory Levels: A Hidden Markov Model Approach

Rainer Schlosser[(✉)] and Keven Richly[(✉)]

Hasso Plattner Institute, University of Potsdam, Potsdam, Germany
{rainer.schlosser,keven.richly}@hpi.de

Abstract. Many markets are characterized by competitive settings and incomplete information. While offer prices of sellers are often observable, the competitors' inventory levels are mutually not observable. In this paper, we study stochastic dynamic pricing models in a finite horizon duopoly model with partial information. To be able to derive effective pricing strategies when the competitor's inventory level is not observable, we use a Hidden Markov Model. Our approach is based on feedback pricing strategies that are optimal, if the competitor's inventory level is observable. Optimized price reactions are balancing two effects: (i) to slightly undercut the competitor's price to sell more items, and (ii) to use high prices to promote a competitor's run-out. For the case that a competitor's strategy is unknown, we derive robust heuristic strategies. Comparing duopolies with different information structures, we find that expected sales results are quite similar as long as the firms' information is symmetric. By evaluating asymmetric pairs of strategies, we study to which extent the value of additional information is affected by the consumers' price sensitivity or the competitors' price response times.

Keywords: Dynamic pricing · Duopoly competition ·
Response strategies · Hidden Markov Model · Asymmetric information

1 Introduction

In real-life applications, firms have to deal with competition and limited information. Sellers are required to choose appropriate pricing decisions to maximize their expected profits. In e-commerce, it has become easy to observe and to change prices. Hence, dynamic pricing strategies that take into account the competitor's strategies will be more and more applied.

However, optimal price reactions are not easy to find. Applications can be found in a variety of domains that involve perishable (e.g., airline tickets, accommodation services, seasonal products) as well as durable goods (e.g., technical devices, natural resources).

In this paper, we study duopoly pricing models in a stochastic dynamic framework. We focus on perishable goods. In our model, sales probabilities are allowed

© Springer Nature Switzerland AG 2019
G. H. Parlier et al. (Eds.): ICORES 2018, CCIS 966, pp. 15–36, 2019.
https://doi.org/10.1007/978-3-030-16035-7_2

to be an arbitrary function of time and the competitor's prices. Our aim is to take into account scenarios in which (i) the competitor's inventory level *is* observable, (ii) the competitor's inventory level *is not* observable, and (iii) even the competitor's pricing strategy is unknown.

1.1 Literature Review

To optimally sell products is a classical application of revenue management theory. The problem is closely related to the field of dynamic pricing, which is summarized in books by Talluri, van Ryzin [1], Phillips [2], and Yeoman, McMahon-Beattie [3]. The survey by Chen, Chen [4] provides an excellent overview of recent pricing models under competition.

Gallego, Wang [5], consider a continuous time multi-product oligopoly for differentiated perishable goods. They use optimality conditions to reduce the multi-dimensional dynamic pure pricing problem to a one dimensional one. Gallego, Hu [6] analyze structural properties of equilibrium strategies in more general oligopoly models for the sale of perishable products. Martinez-de-Albeniz, Talluri [7] consider duopoly and oligopoly pricing models for identical products. They use a general stochastic counting process to model the demand of customers.

Further related models are studied by Yang, Xia [8] and Wu, Wu [9]. Dynamic pricing models under competition that also include strategic customers are analyzed by Levin et al. [10] and Liu, Zhang [11]. Competitive pricing models with limited demand information are studied by Tsai, Hung [12], Adida, Perakis [13], and Chung et al. [14] using robust optimization and demand learning approaches. The effects of strategic interaction of data-driven policies in competitive settings are studied by, e.g., Kephart et al. [15] or Serth et al. [16], using interactive simulation platforms.

In most existing models strong assumptions are made: (i) sales probabilities are assumed to be of a highly stylized form, (ii) the competitors' inventory levels are assumed to be observable, and (iii) competitors adjust their prices at the same point in time. While many papers concentrate on (the existence of) equilibrium strategies, we look for applicable solution algorithms that allow to compute effective response strategies in more realistic settings: Demand probabilities are allowed to generally depend on time as well as prices of all market participants. Inventory levels do not have to be mutually observable. As in practical applications, we assume sequential mutual price reactions with some delay. We consider a discrete time model which is based on the infinite horizon model described in [17]. We extend their model by additional inventory considerations and a finite horizon setting.

1.2 Contribution

This paper is an extended version of [18]. The main contribution of [18] is threefold. We (i) derive optimal pricing strategies when the competitor's inventory level is observable, (ii) derive near-optimal pricing strategies for the case that the competitor's inventory level is not observable, and (iii) we present a heuristic for the case that competitors' strategies are not known.

Compared to [18], in this paper, we present extended evaluation studies and make the following contributions: First, to determine the value of information, we let our three types of strategies play against each other in different duopoly setups. We show that in different symmetric setups sales results are quite similar. Our evaluations of asymmetric strategy setups show that additional information leads to significantly higher profits (compared to the competitor). We also observe that strategies that use more information tend to have higher standard deviations of profits and a lower load factor. Second, we study to which extent performance results of various competitive setups are affected by the consumers' price sensitivity. We find that a higher price sensitivity (e.g., when customers are less loyal) does not lead to a significant decrease in expected profits. Third, we study the impact of price response times on our strategies' performances under various competitive setups. We observe that higher price reaction frequencies can even overcompensate a lack of information.

The remainder of this paper is organized as follows. In Sect. 2, we describe the stochastic dynamic duopoly model for the sale of a finite number of perishable goods. We allow sales intensities to depend on the competitor's price as well as on time (cf. seasonal effects). The state space of our model is characterized by time and the current competitors' prices. The stochastic dynamic control problem is expressed in discrete time.

In Sect. 3, we consider a duopoly competition, in which the inventory level of the competitor is observable. We assume that both competitors act rationally. We set up a firm's Hamilton-Jacobi-Bellman equation and use recursive methods (value iteration) to compute both firms' value functions. Finally, we are able to compute optimal feedback prices as well as expected profits of the two competing firms. By using numerical examples, we investigate typical properties of optimal pricing policies.

In Sect. 4, we analyze response strategies for cases where the inventory level of the competitor is not observable. By using a Hidden Markov Model, we show how to compute efficient pricing strategies and how to evaluate expected profits. Our proposed solution approach is based on the results of the full information model introduced in the previous section. The key idea is to let the competing firms mutually estimate their competitor's remaining inventory level. In Sect. 5, we show how to derive applicable dynamic pricing heuristics for cases in which the competitor's inventory level as well as its pricing strategy are unknown.

In Sect. 6, we compare the different strategies derived in this paper using various numerical experiments. We consider symmetric as well as asymmetric combinations of strategies that use different information structures. Conclusions and future work are given in the final section.

2   Model Description

We consider a situation in which a firm wants to sell a finite number of perishable goods (e.g., airline tickets, hotel tickets, etc.) on a digital market platform.

We assume that a second seller competes for the same market. In our model, we allow customers to compare prices of the two different competitors.

The initial numbers of items of firm 1 and firm 2 are denoted by $N^{(1)}$ and $N^{(2)}$, respectively, $N^{(1)}, N^{(2)} < \infty$. We assume that items cannot be reproduced or reordered. The time horizon T is finite, $T < \infty$. If firm k sells one item, the shipping costs $c^{(k)}$ have to be paid, $k = 1, 2$. A sale of one of firm k's items at price a leads to a net revenue of $a - c^{(k)}$. Discounting is also included in the model. For the length of one period we use the discount factor δ, $0 < \delta \leq 1$.

Due to customer choice the sales probabilities of a firm should depend on its offer price a and the competitor's price p. We also allow the sales probabilities to depend on time.

The (joint) probability that between time t and $t + \Delta$ firm 1 can sell exactly i items at a price a, $a \geq 0$, while firm 2 can sell j items at price p, $p \geq 0$, is denoted by, $0 \leq t < T$, $\Delta > 0$, $i, j = 0, 1, 2, ...,$

$$P_t^{(\Delta)}(i, j, a, p).$$

Without loss of generality, in the following, we assume Poisson distributed sales probabilities, i.e.,

$$
\begin{aligned}
P_t^{(\Delta)}(i, j, a, p) := {} & \frac{\Lambda_{t,\Delta}^{(1)}(a, p)^i}{i!} \cdot e^{-\Lambda_{t,\Delta}^{(1)}(a,p)} \\
& \cdot \frac{\Lambda_{t,\Delta}^{(2)}(p, a)^j}{j!} \cdot e^{-\Lambda_{t,\Delta}^{(2)}(p,a)},
\end{aligned}
\tag{1}
$$

where $\Lambda_{t,\Delta}^{(k)}(a, p) := \int_t^{t+\Delta} \lambda_s^{(k)}(a, p)\, ds$, $k = 1, 2$, $a, p \geq 0$; the sales intensity of a firm k's product is denoted by $\lambda^{(k)}$. In our model, the sales intensity of firm k, $k = 1, 2$, $t \in [0, T]$, $a \geq 0$, $p \geq 0$,

$$\lambda_t^{(k)}(a, p) \tag{2}$$

is a general function of time t, offer price a, and the competitor's price p. The random inventory level of firm k at time t is denoted by $X_t^{(k)}$, $0 \leq t \leq T$. The end of sale for firm k is the random time $\tau^{(k)}$, when all of its items are sold, that is $\tau^{(k)} := \min_{0 \leq t \leq T}\{t : X_t^{(k)} = 0\} \wedge T$; for all remaining $t \geq \tau$ we let a firm's price $a_t := 0$ and $\lambda_t^{(k)}(0, \cdot) := 0$, $k = 1, 2$. As long as a firm has items left to sell, for each period t, a price a has to be chosen.

We call strategies $(a_t)_t$ admissible if they belong to the class of Markovian feedback policies; i.e., pricing decisions $a_t \geq 0$ may depend on time t, the current own inventory level, the current prices of the competitor, and (if observable) the inventory level of the competitor. By A we denote the set of admissible prices. A list of variables and parameters is given in the Appendix, see Table 7.

In some applications, sellers are able to anticipate transitions of the market situation. In particular, the price responses of competitors as well as their reaction time can be taken into account. In this case, a change of the competitor's price p can take place within one period. A typical scenario is that a competitor adjusts its price in response to another competitor's price adjustment with a certain delay.

In the following two sections, we assume that the pricing strategy and the reaction time of competitors are known. We assume that choosing a price a at time t is followed by a state transition (e.g., a competitor's price reaction) and the current price p changes to a subsequent price reaction, which may depend on the current price decision a.

We assume that the state of the system is characterized by the inventory levels of both firms and the current competitor's price. In real-life applications, a firm is not able to adjust its prices immediately after the price reaction of the competing firm. Hence, we assume that in each period the price reaction of the competing firm (firm 2) takes place with a delay of h periods, $0 < h < 1$. After an interval of size h the competitor adjusts its price, see Fig. 1. Firm 1 responds to firm 2 with a delay of $1 - h$.

In period t the probability to sell exactly i items during the first interval of size h, i.e., $[t, t+h]$, is $P_t^{(h)}(i, j, a_t, p_{t-1+h})$, $t = 0, 1, ..., T - 1$. Due to the competitor's price reaction for the rest of the period $[t + h, t + 1]$ the sales probability changes to $P_{t+h}^{(1-h)}(i, j, a_t, p_{t+h})$, $t = 0, 1, ..., T - 1$.

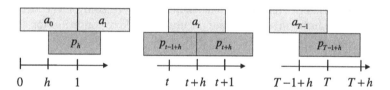

Fig. 1. Sequence of price reactions in a duopoly with reaction time h and $1 - h$, respectively, $0 < h < 1$, cf. [18].

For single intervals $[0, h]$ and $[T, T + h]$, we assume that there is no demand and we let $P_0^{(h)}(i, j, a_0, p_0) = P_T^{(h)}(i, j, a_T, p_{T-1+h}) := 1_{\{i=j=0\}}$.

The evolution of the accumulated profit of firm k, $k = 1, 2$, is connected to its inventory process $X_t^{(k)}$ and characterized by each period's realized net revenues. Depending on the chosen pricing strategy $(a_t)_t$ of firm 1 and the strategy $(p_t)_t$ of firm 2, the random accumulated profit of firm k from time t on (discounted on time t) amounts to, $0 \leq t \leq T$, $k = 1, 2$,

$$G_t^{(k)} := \sum_{s=t}^{T-1} \delta^{s-t} \cdot (a_s - c^{(k)}) \cdot \left(X_s^{(k)} - X_{s+1}^{(k)} \right). \tag{3}$$

Each firm k seeks to determine a non-anticipating (Markovian) pricing policy that maximizes its expected total profit, $k = 1, 2$,

$$E\left(G_0^{(k)} \mid X_0^{(1)} = N^{(1)}, X_0^{(2)} = N^{(2)} \right). \tag{4}$$

In the following sections, we solve dynamic pricing problems that are related to (1)–(4). In the next section, we consider competitive duopoly markets with complete information. In Sect. 4, we compute pricing strategies for scenarios with incomplete information and partially observable states, i.e., we assume that the competitor's inventory level is not observable. In Sect. 5, we additionally assume that the competitor's strategy is unknown. In Sect. 6, we compare the results of the three different models using extensive numerical experiments.

3 Optimal Dynamic Pricing Strategies in a Duopoly with Observable States

In this section, we want to derive mutual optimal price response strategies. We assume that both firms can mutually observe their inventory levels.

3.1 Solution with Full Knowledge

Following the Bellman approach, the best expected future profits of firm 1 and firm 2, i.e., $E(G_t^{(1)}|X_t^{(1)} = n, X_t^{(2)} = m, p_t = p)$ and $E(G_{t+h}^{(2)}|X_{t+h}^{(1)} = n, X_{t+h}^{(2)} = m, a_{t+h} = a)$, respectively, cf. (4), are described by the value functions $V_t^*(n, m, p)$ and $W_{t+h}^*(n, m, a)$, $t = 0, 1, ..., T$. The set of admissible prices A can be continuous or discrete. If either all items are sold or the time is up, no future profits can be made, i.e., the natural boundary condition for the value functions V and W are given by, $n = 0, 1, ..., N^{(1)}$, $m = 0, 1, ..., N^{(2)}$, $a, p \in A$, $t = 0, 1, ..., T - 1$,

$$V_t^*(0, m, p) = 0, \quad and \quad V_T^*(n, m, p) = 0, \tag{5}$$

$$W_{t+h}^*(n, 0, a) = 0, \quad and \quad W_{T+h}^*(n, m, a) = 0. \tag{6}$$

We assume that in case of a run-out a firm sets its price equal to zero for the rest of the time horizon. The Hamilton-Jacobi-Bellman (HJB) equation of firm 1 can be written as, $t = 0, 1, ..., T - 1$, $n = 1, ..., N^{(1)}$, $m = 0, ..., N^{(2)}$, $0 < h < 1$, $a, p \in A$,

$$
\begin{aligned}
V_t^*(n, m, p) = \max_{a \in A} \Bigg\{ &\sum_{i_1, j_1 \geq 0} P_t^{(h)}(i_1, j_1, a, p) \\
&\cdot \sum_{i_2, j_2 \geq 0} P_{t+h}^{(1-h)}\Big(i_2, j_2, 1_{\{n-i_1>0\}} \cdot a, \\
&\quad p_{t+h}^*\big((n-i_1)^+, (m-j_1)^+, 1_{\{n-i_1>0\}} \cdot a\big)\Big) \\
&\cdot \big((a - c^{(1)}) \cdot \min(n, i_1 + i_2)\big) \\
+ \delta \cdot V_{t+1}^* \Big(&(n - i_1 - i_2)^+, (m - j_1 - j_2)^+, 1_{\{m-j_1-j_2>0\}} \\
&\cdot p_{t+h}^*\big((n - i_1)^+, (m - j_1)^+, 1_{\{n-i_1>0\}} \cdot a\big)\Big)\Bigg\}.
\end{aligned}
\tag{7}
$$

Note, (7) mirrors all possible sales scenarios within one period of time and takes the corresponding inventory transitions as well as the anticipated optimal price reactions of the competitor into account.

The HJB of firm 2 is given by, $t = 0, 1, ..., T - 1$, $n = 0, ..., N^{(1)}$, $m = 1, ..., N^{(2)}$, $0 < h < 1$, $a, p \in A$,

$$W_{t+h}^*(n, m, a) = \max_{p \in A} \left\{ \sum_{i_2, j_2 \geq 0} P_{t+h}^{(1-h)}(i_2, j_2, a, p) \right.$$

$$\cdot \sum_{i_1, j_1 \geq 0} P_{t+1}^{(h)}\left(i_1, j_1, \right.$$

$$a_{t+1}^*\left((n - i_1)^+, (m - j_1)^+, 1_{\{m-j_1>0\}} \cdot p\right), 1_{\{m-j_1>0\}} \cdot p\right) \tag{8}$$

$$\cdot \left((p - c^{(2)}) \cdot \min(m, j_1 + j_2)\right.$$

$$+ \delta \cdot W_{t+1+h}^*\left((n - i_1 - i_2)^+, (m - j_1 - j_2)^+, \right.$$

$$\left. \left. 1_{\{n-i_1-i_2>0\}} \cdot a_{t+1}^*\left((n - i_1)^+, (m - j_1)^+, 1_{\{m-j_1>0\}} \cdot p\right)\right)\right)\right\}.$$

The associated prices of both firms are given by the arg max of (7) and (8), respectively, i.e., $n, m > 0$, $t = 0, 1, ..., T - 1$,

$$a_t^*(n, m, p) = \arg\max_{a \in A} \{...\}, \tag{9}$$

$$p_{t+h}^*(n, m, a) = \arg\max_{p \in A} \{...\}. \tag{10}$$

If a firm runs out of inventory, we set the price 0, i.e., for all m, p we let $a_t^*(0, m, p) = 0$ and for all n, a, we let $p_{t+h}^*(n, 0, a) = 0$. The coupled value functions and the optimal feedback policies of the two competing firms can be computed in the following recursive order, cf. (5)–(6):

$$p_{T-1+h}^*(n, m, a), W_{T-1+h}^*(n, m, a) \rightarrow$$
$$a_{T-1}^*(n, m, p), V_{T-1}^*(n, m, p) \rightarrow ...$$
$$... \rightarrow p_h^*(n, m, a), W_h^*(n, m, a) \tag{11}$$
$$\rightarrow a_0^*(n, m, p), V_0^*(n, m, p).$$

3.2   Numerical Examples

To illustrate the approach, cf. (5)–(11), we consider a numerical example.

Example 3.1. We assume a duopoly. Let $T = 50$, $c^{(1)} = c^{(2)} = 10$, $N^{(1)} = N^{(2)} = 10$, $\delta = 1$, $h = 0.5$, and $a \in A := (10, 20, ..., 400)$. We assume Poisson distributed sales probabilities $P_t^{(h)}(i, j, a, p)$, which are determined by $t = 0, h, 1, ..., T$, $k = 1, 2$, $a, p \in A$, cf. (1),

$$\Lambda_{t,h}^{(k)}(a, p) := h \cdot \left(1 - e^{-10^5 \cdot a^{-2.5+t/T}}\right) \cdot \beta(a, p),$$

and the competition factor $\beta(a,p)$, $a, p \in A$,

$$\beta(a,p) := 1_{\{a>0\}} \cdot \frac{p - L \cdot \min(a,p)}{a + p - 2 \cdot L \cdot \min(a,p)}$$

which is characterized by the competition parameter L, $-\infty < L < 1$. Note, the price sensitivity of customers is increasing in L. For the time being, we let $L := 0.8$.

Table 1 illustrates the expected profits of firm 1 for different inventory levels n and different points in time t (for the case that firm 2's price is $p = 100$ and its inventory level is $N^{(2)} = 10$). We observe that the expected future profits are decreasing in time and increasing-decreasing in the number of items left to sell. The optimal expected profits of the second firm have the same characteristics. Compared to firm 1 the total expected profits of firm 2 are slightly larger $(W_h^*(10, 10, a_0^*(10, 10, 0)) = 1769)$.

Table 1. Expected profits $V_t^*(n, 10, 100)$, Example 3.1, cf. [18].

$n \backslash t$	0	10	20	30	40	45
1	363	362	359	348	306	252
2	654	652	640	601	494	368
3	877	872	852	788	628	423
5	1213	1202	1166	1056	782	381
7	1464	1449	1396	1233	737	381
10	1754	1726	1638	1348	723	381

Table 2 illustrates the feedback prices of firm 1 for different competitor's inventory levels m and different prices p (for the case that $t = 20$ and firm 1's inventory level is $N^{(1)} = 10$). We observe that optimal response prices are decreasing-increasing in the competitor's price and decreasing in the competitor's inventory level. I.e., in general, there is an incentive to (slightly) *undercut* the competitor's price.

However, if the competitor has a small price and a small inventory level then it is more advantageous to set *high* prices such that the competitor is likely to sell all of its remaining items, and in turn, our firm becomes a *monopolist* for the rest of the time horizon. If the competitor's inventory level is small, the optimal price can even dominate the monopoly price, cf. $a_{20}^*(10, 0, 0) = 260$ in Table 2.

Remark 3.1

(i) The expected profits are increasing-decreasing in their own inventory level.
(ii) The expected profits are decreasing in the competitor's inventory level.
(iii) If there is no discounting then the expected profits are increasing in the time-to-go.
(iv) The expected profits are increasing-decreasing in the current competitor's price.

Table 2. Expected profits $a_{20}^*(10, m, p)$, Example 3.1, cf. [18].

$p\backslash m$	0	1	2	3	5	7	10
0	260	.	.	.	.	.	.
50	.	400	390	300	220	200	160
100	.	400	390	300	220	200	160
150	.	400	310	300	220	190	140
200	.	400	280	250	190	180	150
250	.	340	260	200	190	180	150
300	.	240	210	200	190	180	150
400	.	220	200	200	190	180	150

Remark 3.2

(i) The optimal prices are not necessarily decreasing in their own inventory level.
(ii) The optimal prices are decreasing in the competitor's inventory level.
(iii) If demand is not increasing in time then the optimal prices are decreasing in the time.
(iv) The optimal prices are decreasing-increasing in the current competitor's price.

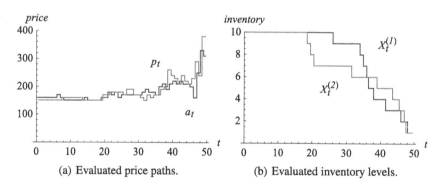

(a) Evaluated price paths.   (b) Evaluated inventory levels.

Fig. 2. Simulated price paths and associated inventory levels over time; Example 3.1, cf. [18].

Figure 2 illustrates simulated sales processes in the context of Example 3.1. Figure 2a illustrates the price trajectories of the two competing firms. Figure 2b shows the associated evolutions of the inventory levels. As demand is increasing in time, on average, prices as well as the number of sales increase at the end of the time horizon.

4 A Hidden Markov Model with Partially Observable States

In this section, we assume that the competitor's inventory level cannot be observed. To derive feedback pricing strategies we use a Hidden Markov Model. We will use probability distributions for the competitor's inventory level, which are based on the observable prices of both firms.

4.1 Theoretical Solution

Let $\pi_t(m)$ denote the (estimated) probability that firm 2 has exactly m items left at time t; let $\varpi_t(n)$ denote the probability that firm 1 has exactly n items left at time t. We assume that the initial inventory levels of both competitors are common knowledge; i.e., the starting distributions are $\pi_0(m) = \pi_h(m) = 1_{\{m=N^{(2)}\}}$ and $\omega_0(n) = \omega_h(n) = 1_{\{n=N^{(1)}\}}$. Furthermore, a run-out is observable, since we assume that in case of a run-out a firm has to set its price equal to zero. The evolutions of the probabilities $\pi_t(m)$ and $\varpi_t(n)$ are given by, $n = 0, ..., N^{(1)}$, $m = 0, ..., N^{(2)}$, $a_t, p_t, a_{t-1+h}, p_{t-1+h} \in A$, $t = 0, 1, ..., T$,

$$\pi_{t+h}(m; a_t, p_t) =$$
$$\sum_{\substack{i_1, j_1 \geq 0, 0 \leq m^- \leq N^{(2)}: \\ m = (m^- - j_1)^+}} P_t^{(h)}(i_1, j_1, a_t, p_t) \cdot \pi_t(m^-)$$

$$\pi_t(m; a_{t-1+h}, p_{t-1+h}) =$$
$$\sum_{\substack{i_2, j_2 \geq 0, \\ 0 \leq m^- \leq N^{(2)}: \\ m = (m^- - j_2)^+}} P_{t-1+h}^{(1-h)}(i_2, j_2, a_{t-1+h}, p_{t-1+h}) \cdot \pi_{t-1+h}(m^-) \tag{12}$$

$$\varpi_{t+h}(n; a_t, p_t) =$$
$$\sum_{\substack{i_1, j_1 \geq 0, 0 \leq n^- \leq N^{(1)}: \\ n = (n^- - i_1)^+}} P_t^{(h)}(i_1, j_1, a_t, p_t) \cdot \varpi_t(n^-)$$

$$\varpi_t(n; a_{t-1+h}, p_{t-1+h}) =$$
$$\sum_{\substack{i_2, j_2 \geq 0, \\ 0 \leq n^- \leq N^{(1)}: \\ n = (n^- - i_2)^+}} P_{t-1+h}^{(1-h)}(i_2, j_2, a_{t-1+h}, p_{t-1+h}) \cdot \varpi_{t-1+h}(n^-). \tag{13}$$

Note, (12) and (13) are relevant for both firms as they might try to estimate (i) the competitor's inventory level as well as (ii) the competitor's beliefs concerning the own inventory. This way the competitor's price reactions can be anticipated via a probability distribution.

Both firms are assumed to act rationally. Pricing decisions are such that no firm has an advantage to deviate from its strategy. Due to the defined sequence of events, theoretically, optimal decisions can be recursively inferred. The corresponding value functions of both firms, denoted by

$$V_t^{(*)}(n, p, \boldsymbol{\pi}_t, \boldsymbol{\omega}_t) \tag{14}$$

$$W_{t+h}^{(*)}(m, a, \boldsymbol{\pi}_{t+h}, \boldsymbol{\omega}_{t+h}), \tag{15}$$

are determined by the usual boundary conditions $V_t^{(*)}(0, \cdot, \cdot, \cdot) = 0$, $V_T^{(*)}(\cdot, \cdot, \cdot, \cdot) = 0$ (for firm 1) and $W_{t+h}^{(*)}(0, \cdot, \cdot, \cdot) = 0$, $W_{T+h}^{(*)}(\cdot, \cdot, \cdot, \cdot) = 0$ (for firm 2) as well as an associated system of Bellman equations similar to (7)–(8) extended by transitions for the beliefs, cf. (12)–(13). The corresponding optimal feedback policies $a_t^{(*)}(n, p, \boldsymbol{\pi}_t, \boldsymbol{\omega}_t)$ and $p_{t+h}^{(*)}(m, a, \boldsymbol{\pi}_{t+h}, \boldsymbol{\omega}_{t+h})$ of the two competing firms can be computed in recursive order (similar to (9)–(11)).

However, optimal policies *cannot* be computed in practical applications. Note, the size of the state space is exploding as the probability distributions $\boldsymbol{\pi}$ and $\boldsymbol{\omega}$ are involved (cf. curse of dimensionality). Hence, heuristic solutions are needed.

In the next subsection, we present an approach to compute viable heuristic feedback pricing strategies for the model with partially observable states. The key idea is to approximate the functions $V_t^{(*)}(n, p, \boldsymbol{\pi}_t, \boldsymbol{\omega}_t)$ and $W_{t+h}^{(*)}(m, a, \boldsymbol{\pi}_{t+h}, \boldsymbol{\omega}_{t+h})$ by using weighted expressions of the value functions $V_t^*(n, m, p)$ and $W_t^*(n, m, a)$ (of the model with full knowledge) and their associated policies $a_t^*(n, m, p)$ and $p_t^*(n, m, a)$ derived in the previous Sect. 3.

4.2   Solution with Partial Knowledge

Motivated by the Hidden Markov Model (HMM), cf. Sect. 4.1, in which the competitor's inventory level cannot be observed, we want to define viable heuristic pricing strategies for the two competing firms. Based on the current beliefs regarding the competitor's inventory, we approximate the correct value functions (14)–(15) (and related controls) using price reactions, cf. (9)–(10), and future profits, cf. (7)–(8), of the fully observable model. As the value functions of the fully observable model might systematically overestimate the correct values (14)–(15), we include an additional positive penalty factor z. If z is smaller than 1, future profits (7)–(8) are reduced.

For firm 1 we define the feedback prices, $t = 0, 1, ..., T - 1$, $n = 1, ..., N^{(1)}$, $p \in A$,

$$\tilde{a}_t(n, p; \boldsymbol{\pi}_t, \boldsymbol{\omega}_t) = \arg\max_{a \in A} \left\{ \sum_{i_1, j_1 \geq 0} P_t^{(h)}(i_1, j_1, a, p) \right.$$

$$\cdot \sum_{0 \leq \tilde{m} \leq N^{(2)}} \pi_t(\tilde{m}) \cdot \sum_{0 \leq \tilde{n} \leq N^{(1)}} \varpi_t(\tilde{n}) \cdot \sum_{i_2, j_2 \geq 0} P_{t+h}^{(1-h)}(i_2, j_2,$$

$$1_{\{\tilde{n}-i_1>0\}} \cdot a, p_{t+h}^* \left((\tilde{n}-i_1)^+, (\tilde{m}-j_1)^+, 1_{\{\tilde{n}-i_1>0\}} \cdot a \right)) \qquad (16)$$

$$\cdot \left((a - c^{(1)}) \cdot \min(n, i_1 + i_2) + \delta \cdot z \right.$$

$$\cdot V_{t+1}^* \left((n - i_1 - i_2)^+, (\tilde{m} - j_1 - j_2)^+, 1_{\{\tilde{m}-j_1-j_2>0\}} \right.$$

$$\left. \left. \left. \cdot p_{t+h}^* \left((\tilde{n}-i_1)^+, (\tilde{m}-j_1)^+, 1_{\{\tilde{n}-i_1>0\}} \cdot a \right) \right) \right) \right\}.$$

Note, (16) mirrors the beliefs for both inventory levels and the corresponding transitions. For anticipated price reactions we use p^*, cf. (10). To estimate future profits, we use $z \cdot V^*$, cf. (7).

Similarly, the prices of firm 2 are given by, $t = 0, 1, ..., T-1$, $m = 1, ..., N^{(2)}$, $a \in A$,

$$\tilde{p}_{t+h}(m, a; \boldsymbol{\pi}_t, \boldsymbol{\omega}_t) = \arg\max_{p \in A} \left\{ \sum_{i_1, j_1 \geq 0} P_{t+h}^{(1-h)}(i_1, j_1, a, p) \right.$$

$$\cdot \sum_{0 \leq \tilde{m} \leq N^{(2)}} \pi_{t+h}(\tilde{m}) \cdot \sum_{0 \leq \tilde{n} \leq N^{(1)}} \varpi_{t+h}(\tilde{n}) \cdot \sum_{i_2, j_2 \geq 0} P_{t+1}^{(h)}(i_2, j_2,$$

$$a_{t+1}^* \left((\tilde{n}-i_1)^+, (\tilde{m}-j_1)^+, 1_{\{\tilde{m}-j_1>0\}} \cdot p \right), 1_{\{\tilde{m}-j_1>0\}} \cdot p) \qquad (17)$$

$$\cdot \left((p - c^{(2)}) \cdot \min(m, j_1 + j_2) + \delta \cdot z \right.$$

$$\cdot W_{t+1+h}^* \left((\tilde{n} - i_1 - i_2)^+, (m - j_1 - j_2)^+, 1_{\{\tilde{n}-i_1-i_2>0\}} \right.$$

$$\left. \left. \left. \cdot a_{t+1}^* \left((\tilde{n}-i_1)^+, (\tilde{m}-j_1)^+, 1_{\{\tilde{m}-j_1>0\}} \cdot p \right) \right) \right) \right\}.$$

In each period, realized sales are used to update the beliefs π and ω such that the prices (16) and (17) can be computed during the sales process, i.e.:

$$\tilde{a}_0(N^{(1)}, 0; \boldsymbol{\pi}_0, \boldsymbol{\omega}_0) \to \boldsymbol{\pi}_h, \boldsymbol{\omega}_h \to \tilde{p}_h(N^{(2)}, a_h; \boldsymbol{\pi}_h, \boldsymbol{\omega}_h)$$

$$\to \boldsymbol{\pi}_1, \boldsymbol{\omega}_1 \to \tilde{a}_1(X_1^{(1)}, p_1; \boldsymbol{\pi}_1, \boldsymbol{\omega}_1) \to ...$$

$$... \tilde{a}_{T-1}(X_{T-1}^{(1)}, p_{T-1}; \boldsymbol{\pi}_{T-1}, \boldsymbol{\omega}_{T-1}) \to \boldsymbol{\pi}_{T-1+h}, \boldsymbol{\omega}_{T-1+h} \qquad (18)$$

$$\to \tilde{p}_{T-1+h}(X_{T-1+h}^{(2)}, a_{T-1+h}; \boldsymbol{\pi}_{T-1+h}, \boldsymbol{\omega}_{T-1+h}).$$

By using simulations both firms' expected profits as well as their distributions can be easily approximated. Evaluating different z values makes it possible to identify the (mutual) best z value.

4.3 Numerical Example

To illustrate our approach, in this subsection, we consider a numerical example.

Example 4.1. We assume the setting of Example 3.1. Both firms use the heuristic Hidden Markov strategies, cf. (16)–(18), for different parameter values z, $0.2 \leq z \leq 1.5$.

We observe that z has an impact on the expected profits of both competing firms. In our example, simulated average profits of both firms are maximized for $z = 0.8$. Note, the lower z is the more risk averse are the pricing policies (see standard deviations σ), cf. Table 3.

Table 3. Simulated expected profits and its standard deviations of both firms for different z values, Example 4.1, cf. [18].

z	$EG_0^{(1)}$	$EG_0^{(2)}$	$EX_T^{(1)}$	$EX_T^{(2)}$	$\sigma(G_0^{(1)})$	$\sigma(G_0^{(2)})$
0.2	1141	1104	0.00	0.00	209	188
0.5	1679	1701	0.44	0.42	249	258
0.6	1743	1741	0.70	0.57	320	283
0.7	1742	1756	0.89	0.79	351	338
0.8	1739	1770	1.15	0.90	397	359
0.9	1732	1753	1.19	1.29	393	420
1.0	1716	1748	1.43	1.40	419	426
1.1	1686	1740	1.72	1.39	452	417
1.2	1668	1715	1.90	1.59	456	427
1.5	1647	1639	2.07	2.31	454	470

Remark 4.1 (Parallelization). The computation of feedback policies and particularly extensive simulation studies can become CPU-intensive. Parallelization can be used to compute results more efficiently:

(i) Feedback prices for the same point in time can run in parallel.
(ii) Simulations can be computed independent from each other.

Figure 3 illustrates simulated sales processes in the context of Example 4.1. Figure 3a illustrates price trajectories of the two competing firms. Figure 3b shows the associated evolutions of the inventory levels and the (mutually) estimated inventory levels of the competitor (dashed lines).

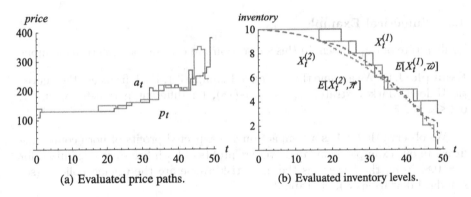

Fig. 3. Simulated price paths and associated (estimated) inventory levels over time, $z = 0.8$; Example 4.1, cf. [18].

5   Unknown Strategies

In this section, we want to present another heuristic approach to derive effective pricing strategies in competitive markets with limited information. We assume that the strategy of the competitor is completely unknown.

Our key idea to deal with unknown price reactions is to assume sticky prices. For firm 1, we define the following value function, $p \in A$, $n \geq 1$, $t = 0, 1, ..., T-1$, $\bar{V}_t(0, p) = 0$ for all t, p, $\bar{V}_T(n, p) = 0$ for all n, p,

$$\bar{V}_t(n, p) = \max_{a \in A} \left\{ \sum_{i_1, j_1} P_t^{(h)}(i_1, j_1, a, p) \right.$$

$$\cdot \sum_{i_2, j_2} P_{t+h}^{(1-h)}(i_2, j_2, a, p) \cdot \left((a - c^{(1)}) \cdot \min(n, i_1 + i_2) \right. \tag{19}$$

$$\left. \left. + \delta \cdot \bar{V}_{t+1} \left((n - i_1 - i_2)^+, p \right) \right) \right\}.$$

The heuristic strategy $\bar{a}_t(n, p)$ – determined by the arg max of (19) – only depends on t, n, and p. Similarly, the corresponding pricing strategy $\bar{p}_t(m, a)$ of firm 2 is determined by the arg max of, $a \in A$, $m \geq 1$, $t = 0, 1, ..., T - 1$, $\bar{W}_{t+h}(0, a) = 0$ for all t, a, $\bar{W}_{T+h}(m, a) = 0$ for all m, a,

$$\bar{W}_{t+h}(m, a) = \max_{p \in A} \left\{ \sum_{i_2, j_2} P_{t+h}^{(1-h)}(i_2, j_2, a, p) \right.$$

$$\cdot \sum_{i_1, j_1} P_{t+1}^{(h)}(i_1, j_1, a, p) \cdot \left((p - c^{(2)}) \cdot \min(m, j_1 + j_2) \right. \tag{20}$$

$$\left. \left. + \delta \cdot \bar{W}_{t+1+h} \left((m - j_1 - j_2)^+, a \right) \right) \right\}.$$

The advantage of this approach is that the value function does not need to be computed for all competitors' prices p in advance. The value function and the associated pricing policy can be computed separately for single prices p (e.g., just when they occur). If the competitor's strategy is not known (which is often the case) it is not possible to anticipate potential price adjustments. This feedback strategy is able to react immediately if a change of the competitor's price takes place. In such an event, the value functions (19)–(20) and the associated prices have to be computed for the new state.

Remark 5.1 (Oligopoly competition). Note, due to the curse of dimensionality, the strategies derived in Sects. 3 and 4 are just applicable when the number of competitors is small. The heuristic strategy described above, however, can still be applied if the number of competitors is large. In case of K competitors, the state p in (19) just have to be replaced by $\boldsymbol{p} = (p^{(1)}, ..., p^{(K)})$, $p^{(k)} \in A$, $k = 1, ..., K$.

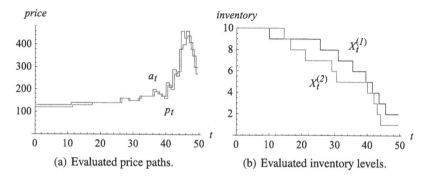

(a) Evaluated price paths. (b) Evaluated inventory levels.

Fig. 4. Simulated price paths and associated inventory levels over time; setting of Example 3.1, cf. [18].

For the case that the competitor's strategy is unknown, Fig. 4 illustrates simulated sales processes based on the heuristic, cf. (19)–(20), in the context of Example 3.1. Figure 4a illustrates price trajectories of the two competing firms. We observe that firms either significantly raise the price or undercut the competitor's price. Figure 4b shows corresponding inventory levels.

6 Evaluation

In this section, we want to compare the outcome of our different solution strategies, which take advantage of different kind of information.

6.1 Comparison of Strategies

If pricing strategies are allowed to use full information (i.e., the own inventory level, the competitor's inventory level, and the competitor's price), the optimal expected profits can be computed analytically, cf. Sect. 3. In case the competitor's inventory level is not known, we presented an approach to compute viable strategies via a Hidden Markov Model, cf. Sect. 4. If the competitor's inventory is not known and her pricing strategy as well as her reaction time is unknown, we proposed an efficient heuristic.

By S_{FK}, we denote the strategy derived in Sect. 3 (full knowledge). By S_{PK}, we denote the response strategy derived in Sect. 4 (partial knowledge) with $z = 0.8$. By S_{UK}, we denote the heuristic strategy, cf. Sect. 5, in case that the competitor's strategy is unknown.

Considering the setting of Examples 3.1 and 4.1, the expected profits of the different symmetric strategy combinations are summarized in Table 4. In all cases, the expected total profits, the expected remaining inventory, and the standard deviations of total profits have been derived using simulations.

Table 4. Strategy comparison (benchmark case $h = 0.5$, $L = 0.8$): Expected profits $EG_0^{(1)}$ (of firm 1) and $EG_0^{(2)}$ (of firm 2), when firm 1 and firm 2 play different pairs of strategies using S_{FK} (full knowledge), S_{PK} (partial knowledge), and S_{UK} (unknown strategies), cf. Examples 3.1–4.1.

Scenario	$EG_0^{(1)}$	$EG_0^{(2)}$	$EX_T^{(1)}$	$EX_T^{(2)}$	$\sigma(G_0^{(1)})$	$\sigma(G_0^{(2)})$
FK vs. FK	1746	1764	1.55	1.52	470	461
PK vs. PK	1739	1770	1.15	0.90	397	359
UK vs. UK	1694	1696	0.37	0.37	373	374
FK vs. PK	1760	1702	1.40	1.13	493	399
PK vs. FK	1723	1810	0.45	2.16	269	588
FK vs. UK	1747	1733	2.16	0.49	574	326
UK vs. FK	1704	1732	0.57	2.07	350	576
PK vs. UK	1721	1603	0.77	0.56	413	331
UK vs. PK	1714	1733	0.40	1.10	305	429

In the first three cases, we observe that in all three *symmetric* scenarios both firms can expect similar results, cf. Figs. 5, 6 and 7. It turns out that as long as the information structure is identical, a lack of information does not necessarily result in smaller expected profits.

The number of unsold items as well as the variance of profits, however, have significant differences. In case of fully observable states (S_{FK} vs. S_{FK}) the remaining inventory and the variance of profits is comparably high. Both firms can expect almost equal results. In the second case with partially observable states (S_{PK} vs. S_{PK}) we observe that the load factor of both firms is higher and the variation of profits is much smaller. Since less information is available the competition between both firms is less intense.

In case of mutual unknown strategies (S_{UK} vs. S_{UK}), we obtain similar results. Furthermore, we can assume that the heuristic strategy S_{UK} will yield robust results when played against various other strategies. The other two strategies are optimized to play against a specific strategy. Hence, they might perform

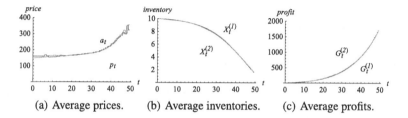

Fig. 5. Simulated expected price paths, associated inventory levels, and accumulated profits over time, full knowledge FK vs. FK; setting of Examples 3.1–4.1.

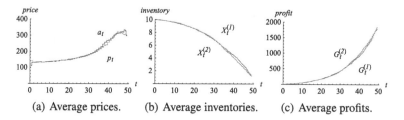

Fig. 6. Simulated expected price paths, associated inventory levels, and accumulated profits over time, partial knowledge PK vs. PK; setting of Examples 3.1–4.1.

less well, when the competitor is playing a different strategy. Moreover, the efficient computation of our heuristic S_{UK} allows fast computation times and, in turn, a high price reaction frequency, which is also a competitive advantage.

In the remaining cases of Table 4, we present the results of asymmetric strategy pairs. As expected, we observe that strategies that have or use more information beat strategies with less information. However, profit differences are relatively small, which means that our strategies with incomplete information are surprisingly competitive.

Further, the firm that has the final price adjustment (firm 2) has a slight advantage. In general, we observe that strategies that use more information tend to have higher standard deviations of profits and a lower load factor.

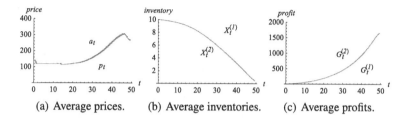

Fig. 7. Simulated expected price paths, associated inventory levels, and accumulated profits over time, no knowledge UK vs. UK; setting of Examples 3.1–4.1.

Note, in the asymmetric setups, both strategies are not optimal response strategies; they are optimized to be played against their symmetric counterpart. Hence, theoretically results could be worse (compared to the symmetric cases) since in our asymmetric setups the competitor might not react as expected. However, we observe that profits are hardly lower. The reason is that the derived strategies (S_{FK}, S_{PK}, S_{UK}) are quite robust due to their feedback nature. Further, in asymmetric setups the competition is less fierce as price reactions are not optimized to be played against the competitor's strategy. For optimized response strategies against given strategies, see [17].

6.2 Impact of Customers Price Sensitivity

In this subsection, we study to which extent results, cf. Table 4, are affected if customers are more price sensitive. Such cases can be modelled using a higher competition factor L, cf. Example 3.1. Similarly, a lower factor L corresponds to cases in which customers are more loyal and tend to stick to a certain firm instead of steadily comparing prices.

Table 5 summarizes the performance results for all symmetric and asymmetric duopoly scenarios for the case $L := 0.95$. Again, results were computed using simulation studies.

In case of a higher price sensitivity, we again observe that strategies are more successful if more information is used/available. More interestingly, we observe that (compared to the benchmark case, cf. Table 4) due to fierce competition it is more important whether a firm has the last move. One might think that in cases with high price sensitivity profits are lower as products with the higher price can hardly be sold, and in turn, both firms are forced to systematically undercut the competitor's price in order to sell items (race to the bottom). Surprisingly profits are not necessarily lower! The reason is that the effects of a higher price sensitivity are counterbalanced by the fact that the firm, which sells less fast is likely to become a monopolist for the rest of the time horizon.

6.3 Impact of Reaction Time

In this subsection, we investigate the impact of reaction times on our strategies' performance results. In our model the reaction time can be varied via the parameter h, $0 < h < 1$. While firm 2 reacts on firm 1's action with a delay of h, firm 1's reaction time on firm 2's price adjustment is $1 - h$. A reaction time $h = 0.2$ corresponds to the case in which firm 1 has $h = 20\%$ of the time the "fresh" price; firm 2's share is $1 - h = 80\%$.

In real-life applications, firms often randomize their reaction time in order not to act predictably. In this case, the ratio of the competing firms' reaction frequencies determines the share of time a firm has the most recent price update. In [17] it is demonstrated that such scenarios can be effectively modelled via our duopoly model with fixed reaction times h and $1 - h$, respectively.

Table 5. Impact of price sensitivity factor L (case $h = 0.5$, $L = 0.95$): Expected profits $EG_0^{(1)}$ (of firm 1) and $EG_0^{(2)}$ (of firm 2), when firm 1 and firm 2 play different pairs of strategies using S_{FK} (full knowledge), S_{PK} (partial knowledge), and S_{UK} (unknown strategies), cf. Example 3.1–4.1.

Scenario	$EG_0^{(1)}$	$EG_0^{(2)}$	$EX_T^{(1)}$	$EX_T^{(2)}$	$\sigma(G_0^{(1)})$	$\sigma(G_0^{(2)})$
FK vs. FK	1784	1795	1.14	2.15	429	556
PK vs. PK	1696	1724	0.90	0.60	352	297
UK vs. UK	1574	1575	0.27	0.28	414	417
FK vs. PK	1757	1743	1.80	0.60	504	282
PK vs. FK	1639	1811	1.40	1.00	405	448
FK vs. UK	1741	1732	2.23	0.36	590	300
UK vs. FK	1697	1724	0.45	2.16	331	589
PK vs. UK	1716	1660	1.00	0.30	403	268
UK vs. PK	1658	1719	0.40	1.20	269	450

Table 6. Impact of reaction time (case $h = 0.2$ vs. $h = 0.8$, $L = 0.8$): Expected profits $EG_0^{(1)}$ (of firm 1) and $EG_0^{(2)}$ (of firm 2), when firm 1 and firm 2 play different pairs of strategies using S_{FK} (full knowledge), S_{PK} (partial knowledge), and S_{UK} (unknown strategies), cf. Examples 3.1–4.1.

Scenario	h	$EG_0^{(1)}$	$EG_0^{(2)}$	$EX_T^{(1)}$	$EX_T^{(2)}$	$\sigma(G_0^{(1)})$	$\sigma(G_0^{(2)})$
FK vs. FK	0.2	1734	1786	0.97	1.95	389	505
FK vs. FK	0.8	1782	1732	1.98	0.93	509	381
PK vs. PK	0.2	1675	1860	1.40	0.90	410	362
PK vs. PK	0.8	1858	1703	0.60	1.20	304	412
UK vs. UK	0.2	1677	1715	0.37	0.33	362	366
UK vs. UK	0.8	1712	1674	0.33	0.37	372	363
FK vs. PK	0.2	1730	1716	0.70	1.41	396	460
FK vs. PK	0.8	1876	1633	0.90	1.50	400	404
PK vs. FK	0.2	1616	1881	1.80	0.80	455	382
PK vs. FK	0.8	1749	1776	0.60	1.50	313	525
FK vs. UK	0.2	1721	1746	2.07	0.41	336	236
FK vs. UK	0.8	1724	1659	1.85	0.61	545	372
UK vs. FK	0.2	1648	1724	0.62	1.77	374	545
UK vs. FK	0.8	1741	1728	0.41	2.08	301	532
PK vs. UK	0.2	1758	1779	1.00	0.40	376	273
PK vs. UK	0.8	1744	1691	1.40	0.50	497	319
UK vs. PK	0.2	1627	1705	0.60	1.10	325	430
UK vs. PK	0.8	1722	1666	0.50	1.30	254	409

To this end, Table 6 shows simulated performance results for all duopoly scenarios for two different (uneven) reaction times $h = 0.2$ and $h = 0.8$. The price sensitivity factor is $L = 0.8$.

We observe that, in general, profits are significantly affected by response times. Hence, price update frequencies are a competitive advantage. We find that the competitor with a better (more frequent) reaction time can even beat its opponent although a strategy with using less information is applied, i.e., a better reaction time can overcompensate the lack of information.

7   Conclusion

In e-commerce, it has become easier to observe and adjust prices automatically. Consequently, there exists an increased demand for dynamic pricing. The computation of suitable pricing strategies is highly challenging as soon as strategic competitors are involved and remaining inventory levels play a major role. In this paper, we analyzed stochastic dynamic finite horizon duopoly models characterized by price responses in discrete time. We allow sales probabilities to generally depend on time as well as the competitors' prices. Further, we are able to model different price reaction times.

We have considered three different types of information structures. In the first setting, we assume that the inventory levels of the competing firms are mutually observable. We show that optimal price reaction strategies – which are based on mutual price anticipations – can be derived using standard methods (e.g., backward induction). Examples are used to identify structural properties of expected profits and feedback pricing strategies. Optimal prices are balancing two effects: (i) slightly undercut the competitor's price in order to sell more items, and (ii) the use of high prices in order to promote a competitor's run-out and to act as a monopolist for the rest of the time horizon.

In the second setting, we assume that the inventory of the competitor is not observable. Based on observable prices, we compute probability distributions (beliefs) for the number of items the competitor might have left to sell. We propose a Hidden Markov Model to be able to compute applicable feedback pricing strategies. Our examples show that the resulting expected profits of both firms are similar to those obtained in the model with full knowledge. The variance of profits and the average number of remaining items, however, is significantly lower.

In the third setting, we assume that the competitor's strategy is completely unknown, i.e., competitors cannot anticipate price responses. We propose an efficient decomposition approach to circumvent the curse of dimensionality and demonstrate how to compute powerful pricing strategies. We verify that – when applied by both competitors – the heuristic yields the same expected profits as in the two other settings, in which more information is available.

We have shown how to compute applicable reaction strategies for real-life scenarios with different information structures. We find that sales results are quite similar as long as the information structure is symmetric. Our numerical experiments of asymmetric strategy setups show that additional information leads to significantly higher profits (compared to the competitor). Further, we observe that a higher price sensitivity (e.g., when customers are less loyal) does not lead to a significant decrease in expected profits. Moreover, we find that higher price reaction frequencies can even overcompensate a lack of information.

In future research, the model could be extended to study scenarios with (i) multiple products and substitution effects in demand, (ii) strategic customers that anticipate typical price trends, or (iii) competitors that seek to learn the competitors' pricing strategy based on historic data.

Appendix

Table 7. Notation table.

t	Time/Period
T	Time horizon
$c^{(k)}$	Shipping costs of firm k, $k = 1, 2$
$G_t^{(k)}$	Random future profits of firm k
$N_t^{(k)}$	Initial number of sold items of firm k
$X_t^{(k)}$	Random inventory level of firm k
δ	Discount factor
h	Reaction time (of firm 2)
$P_t^{(h)}$	Sales probability for $(t, t + h)$
β	Competition factor
L	Price sensitivity factor
A	Set of admissible prices
V	Value function of firm 1
W	Value function of firm 2
a	Offer price of firm 1
p	Offer price of firm 2
n	Inventory state of firm 1
m	Inventory state of firm 2
$\pi(m)$	Beliefs of firm 1
$\omega(n)$	Beliefs of firm 2
a^*, p^*	Feedback prices (full knowledge model)
$\tilde{a}, \tilde{p}$	Feedback prices (partial knowledge model)
$\bar{a}, \bar{p}$	Feedback prices (no knowledge model)
FK	Full knowledge
PK	Partial knowledge
UK	No knowledge (unknown)

References

1. Talluri, K.T., Van Ryzin, G.J.: The Theory and Practice of Revenue Management, vol. 68. Springer, New York (2004). https://doi.org/10.1007/b139000
2. Phillips, R.L.: Pricing and Revenue Optimization. Stanford University Press, Stanford (2005)
3. Yeoman, I., McMahon-Beattie, U. (eds.): Revenue Management: A Practical Pricing Perspective. Palgrave Macmillan, London (2011). https://doi.org/10.1057/9780230294776
4. Chen, M., Chen, Z.L.: Recent developments in dynamic pricing research: multiple products, competition, and limited demand information. Prod. Oper. Manag. **24**, 704–731 (2015)
5. Gallego, G., Wang, R.: Multiproduct price optimization and competition under the nested logit model with product-differentiated price sensitivities. Oper. Res. **62**, 450–461 (2014)
6. Gallego, G., Hu, M.: Dynamic pricing of perishable assets under competition. Manage. Sci. **60**, 1241–1259 (2014)
7. Martínez-de Albéniz, V., Talluri, K.: Dynamic price competition with fixed capacities. Manage. Sci. **57**, 1078–1093 (2011)
8. Yang, J., Xia, Y.: A nonatomic-game approach to dynamic pricing under competition. Prod. Oper. Manag. **22**, 88–103 (2013)
9. Wu, L.L.B., Wu, D.: Dynamic pricing and risk analytics under competition and stochastic reference price effects. IEEE Trans. Industr. Inf. **12**, 1282–1293 (2016)
10. Levin, Y., McGill, J., Nediak, M.: Dynamic pricing in the presence of strategic consumers and oligopolistic competition. Manage. Sci. **55**, 32–46 (2009)
11. Liu, Q., Zhang, D.: Dynamic pricing competition with strategic customers under vertical product differentiation. Manage. Sci. **59**, 84–101 (2013)
12. Tsai, W.H., Hung, S.J.: Dynamic pricing and revenue management process in internet retailing under uncertainty: an integrated real options approach. Omega **37**, 471–481 (2009)
13. Adida, E., Perakis, G.: Dynamic pricing and inventory control: uncertainty and competition. Oper. Res. **58**, 289–302 (2010)
14. Chung, B.D., Li, J., Yao, T., Kwon, C., Friesz, T.L.: Demand learning and dynamic pricing under competition in a state-space framework. IEEE Trans. Eng. Manage. **59**, 240–249 (2012)
15. Kephart, J.O., Hanson, J.E., Greenwald, A.R.: Dynamic pricing by software agents. Comput. Netw. **32**, 731–752 (2000)
16. Serth, S., et al.: An interactive platform to simulate dynamic pricing competition on online marketplaces. In: 21st IEEE International Enterprise Distributed Object Computing Conference, EDOC, pp. 61–66 (2017)
17. Schlosser, R., Boissier, M.: Optimal price reaction strategies in the presence of active and passive competitors. In: International Conference on Operations Research and Enterprise Systems, pp. 47–56 (2017)
18. Schlosser, R., Richly, K.: Dynamic pricing strategies in a finite horizon duopoly with partial information. In: International Conference on Operations Research and Enterprise Systems, pp. 21–30 (2018)

A Bayesian Inference Analysis of Supply Chain Enablers, Supply Chain Management Practices, and Performance

Behnam Azhdari[✉]

Department of Management, Islamic Azad University, Khark Branch,
Shohada St., Khark Island, Boushehr, Iran
bajdari@ut.ac.ir

Abstract. In this study, a Causal Bayesian network (CBN) model of the causal relationships between supply chain enablers, supply chain management practices and supply chain performances is empirically developed and analyzed. Study data collected from a sample of 199 manufacturing firms producing the most influential products in Iran's economy. Resultant CBN model revealed important causalities between study variables of interest. Afterwards, using Dirichlet estimator of TETRAD 6-4-0 software, conditional probability estimation with Bayesian networks, also known as Bayesian inference was developed. The outcomes of this study in general, support the idea that SC enablers, especially IT technologies, don't have direct impact on SC performance. Also forward Bayesian inference provided deeper understanding of causal relationships in supply chain context, such as what antecedents must be available to reach better level at each critical supply chain performance measures. Also it is found out that in any tier of supply chain concepts; there may be some important intra-relations which worth of further studies.

Keywords: Supply chain management · Supply chain performance ·
Causal Bayesian network · Bayesian inference

1 Introduction

Today's business competition is mostly among supply chains and not just between individual organizations. Supply chain (SC) enablers are required tools to practice effective supply chain management. So, to improve SC performance, it is necessary to study the impact of SC enablers and SCM practices on SC performance. As posited by Hsu et al. [1], effective supply chain management practices are vital antecedents of supply chain competitive advantage and performance. The existing literature provides numerous examples of companies that have gained a competitive advantage by using superior supply chain management practices [2]. As stated by Li et al. [3] despite the importance of implementing SCM practices, organizations often do not know exactly what to implement, due to a lack of understanding of what constitutes a comprehensive set of SCM practices. In addition, organizations don't know how practically can increase their supply chain performance through these practices and what enablers are exactly needed.

© Springer Nature Switzerland AG 2019
G. H. Parlier et al. (Eds.): ICORES 2018, CCIS 966, pp. 37–53, 2019.
https://doi.org/10.1007/978-3-030-16035-7_3

The goal of this research is to develop a causal Bayesian network (CBN) model of the relations between SC enablers, SCM practices and SC performance in supply chain and then to analyze its conditional probabilities by means of Bayesian inference. The reminder of this paper is as follows. In Sect. 2, influential papers about relationships between SC enablers, SCM practices and performance reviewed. Then, the data collection and measurement model development are discussed in Sect. 3. In Sect. 4, causal Bayesian network development and Bayesian inference analysis is presented. In Sect. 5, the results and implications are deliberated. Conclusions and study limitations and also future research suggestions are discussed in Sect. 6.

2 Theoretical Background

2.1 Relationships Between SC Enablers, SCM Practices and SC Performance

Studying the relationships between SC enablers and SCM practices and their effect on performance is interesting to many academics and SCM practitioners. A review of these works is presented in [4] which depicted in Table 1. As this table shows, the authors of these studies were more focused on organizational performance [5–8].

In one of the first papers in this context that considers SC performance, Shin et al. [9] worked on the effect of supply chain management orientations on SC performance. They concluded that improvement in supply chain management orientation, including some SC practices, can improve both the suppliers' and buyers' performance. In other study, Lockamy and McCormack [10] investigated the relationships between SCOR model planning practices with SC performance. They reported that planning processes are critical in all SCOR supply chain planning decision areas and collaboration is the most important factor in the plan, source and make planning decision areas. Lee et al. [11] also studied the relationships between three SC practices, including supplier linkage, internal linkage and customer linkage, and SC performance. They concluded that internal linkage is a main factor of cost-containment performance and supplier linkage is a crucial indicator of performance reliability as well as performance. In another work, Sezen [12] investigated the relative effects of three SCM practices including supply chain integration, supply chain information sharing and supply chain design on supply chain performance. He concluded that the most important effect on resource and output performances belongs to supply chain design. He also concluded that information sharing and integration are correlated with performance, but their effect strength are lower than supply chain design. In one of the newest works in this area, Ibrahim and Ogunyemi [13] tested the effect of information sharing and supply chain linkages on supply chain performance. Their results reveal that supply chain linkages and information sharing, positively related to flexibility and efficiency of supply chain.

Seemingly the first article, in which authors consider the effects of both SC enablers and SCM practices on SC performance, is the study of Li et al. [14]. They investigated the relations between three factors including IT implementation as an important SC enabler, supply chain integration as an SCM practice, and SC performance. As a result, they suggested that IT implementation has no direct impact on SC performance, but it improves SC performance through its positive impact on SC integration. In other work, Zelbst et al. [15] theorized and assessed a structural model that includes RFID technology utilization and supply chain information sharing as antecedents to supply chain performance. The results of their work show that although RFID technology does not directly influence on SC performance, its utilization leads to improve information sharing among supply chain members, which in turn leads to improve SC performance.

Table 1. Relationships between SC enablers, SCM practices and SC performance in the literature [4].

References	Scope of SC enablers	Scope of SCM practices	Methodology	Scope of performance measurement
Narasimhan and Jayanth [5]	–	Narrow	SEM[a]	Organization
Shin et al. [9]	–	Narrow	SEM	Supply chain
Frohlich and Westbrook [6]	–	Narrow	ANOVA[b]	Organization
Tan et al. [7]	–	Wide	Correlation	Organization
Lockamy III and McCormack [10]	–	Narrow	Regression	Supply chain
Li and Lin [8]	Wide	Wide	Regression	–
Li et al. [3]	–	Wide	SEM	Organization
González-Benito [16]	Narrow	Narrow	SEM	Organization
Sanders [17]	Narrow	Narrow	SEM	Organization
Zhou and Benton Jr. [18]	Narrow	Narrow	SEM	–
Li et al. [19]	–	Narrow	SEM	Organization
Lee et al. [11]	–	Narrow	Multiple regression	Supply chain
Johnson et al. [20]	Wide	–	Regression	Organization
Devaraj et al. [21]	Narrow	Narrow	SEM	Organization
Sezen [12]	–	Narrow	Regression	Supply chain
Li et al. [14]	Wide	Narrow	SEM	Supply chain
Bayraktar et al. [22]	–	Wide	SEM	Organization
Hsu [1]	–	Wide	SEM	Organization
Davis-Sramek et al. [23]	Narrow	–	Regression	Organization
Zelbst et al. [15]	Narrow	Narrow	SEM	Supply chain
Sundram et al. [24]	–	Wide	PLS[c]	Supply chain
Hamister [25]	–	Wide	PLS	Supply chain
Ibrahim and Ogunyemi [13]	–	Narrow	Regression	Supply chain

[a]Structural Equation Modeling
[b]Analysis of variance
[c]Partial Least Squares

2.2 Bayesian Inference in Supply Chain Management Studies

There is scarce papers which focus on Bayesian inference in supply chain management. Ding et al. [26] in their paper, used Bayesian networks to model dependencies between managed objects in distributed systems and backward inference to fault locating in supply chain. In the other work, Antai [27], suggested a conceptualization of supply chain versus supply chain competition using the Bayesian inference approach by simulated data. Markis et al. [28] in their paper presented a Bayesian inference method of quantifying a buyer's likelihood to purchase a highly customized product in automotive industry. In the last reviewed paper, Garvey et al. [29] utilized a Bayesian network approach to risk propagation in a supply network, taking into account the inter-dependencies among different risks, as well as the idiosyncrasies of a supply chain network structure.

Fig. 1. The proposed basic conceptual model [4].

2.3 Conceptual Model

Although there is no doubt about the importance of the relations between SC enablers, SCM practices and SC performance, not many studies can be found in the literature which cover these relations in a whole model. Thus, in this research a basic conceptual model of relationships among SC enablers, SCM practices and SC performance developed (Fig. 1). As depicted in this model, based on the literature [15, 30] this research suggests that SC enablers have direct impact on SCM practices and no direct impact on SC performance.

3 Research Methodology

3.1 Questionnaire

After a comprehensive supply chain management literature review, 20 articles that indicate SCM practices or activities and 10 articles that indicate SC enablers have been considered. Then 54 practices and 22 enablers cited in these articles were identified.

In order to achieve a valid list of SC enablers and SCM practices to include in the questionnaire, Q-sort methodology was used. To apply Q-sort method, six researchers and experts were asked to classify the specified initial items into SC enabler and SCM practice categories. Q-sort resulted in 20 SC enablers out of 22 and 44 SCM practices out of 54 initial items. The judges' agreement for these items was more than 70%, which is above the recommended value of 65% [31]. Towards a final list of SC enablers and SCM practices, content analysis was used to identify similar statements and merge some similar items to definitive ones. As a result, 7 SC enablers and 8 SCM practices were identified and they are shown in Table 2. In case of SCM practices the respondents were asked to indicate that what extent these scale items were implemented in SCM of their core products, relying on five-point scales ranging from 1 = 'not at all implemented' to 5 = 'fully implemented'. In case of SC enablers, the respondents were asked to indicate their perceptions of relative importance of these enablers in SCM of their core products on five-point scales ranging from 1 = 'of no importance' to 5 = 'of major importance'.

To identify important SC performance measures, supply chain management processes of SCOR model was used, including scale items for measuring 'SCM planning', 'logistics performance', 'supply chain production performance', 'supply chain delivery performance', and 'customer delight performance'. The respondents were asked to indicate on a 6-point scale, ranging from 1 = 'definitely worse' to 6 = 'definitely better', on how their core products supply chain had performed relative to their major competitors or their overall industry on each of these supply chain performance criteria.

Table 2. Final SC enablers and SCM practices [4].

	Survey constructs
SC enabler	e-supply chain portal
	Performance measurement systems
	Advanced manufacturing technology
	Inter-organizational communication technology
	Logistic infrastructure
	e-commerce technologies
	Unique identification and trace technologies
SCM practices	Information sharing
	Strategic view in supply chain management
	Lean manufacturing practices
	Supplier management
	Performance management
	Human resources management
	Customer orientation
	Supply chain integration

3.2 Data Collection

Before data collection, a panel of 4 researchers' were asked to evaluate the questionnaire, regarding ambiguity, appropriateness, and completeness. By reviewing a few resulted comments, the survey questionnaire was modified and finalized.

Target sample of study was collected from manufacturers of 10 products classes, covered by IranCode® products classification system. These products are the most influential in Iranian economy. It was suggested that the firms with more products have more structured supply chain so more suitable to be included in the sample of this study. Herein the firms were sorted, based on the number of their registered products in

IranCode®. Then, using stratified random sampling, a group of 2000 firms was selected and were asked to fill out the questionnaire. After four weeks, as follow up procedure, personalized reminder e-mails were sent to potential participants. Finally, out of 2000 surveys mailed, 199 valid responses were received, resulting in a response rate of 11.63%, which is acceptable as some other studies in this field [8, 32].

Non-response bias measured by applying a t-test on the scores of early and late responses. The responses were divided into two groups: 142 responses (71.4%) received within 3 weeks after mailing, and 57 ones (28.6%) received four weeks later and even more. The result of this test indicated no significant difference between the two groups.

As this study based on single respondents and perceptual scales, the risk of common method variance was assessed, so a model was run without the method factor and it was compared to the one with method factor added [32]. Since the method factor failed to change substantive conclusions, it was concluded that the amount and extent of method variance does not harm the validity of the measurement model.

Sample responses included 24% food products manufacturers, 19.8% road making machinery and construction materials manufacturers, 12.8% chemical manufacturers, 11.2% medical and cosmetic manufacturers, 9.6% industries general necessities manufacturers, 8.6% auto parts manufacturers and 13.8% other manufacturers. Of all respondents, 28% were CEO, President, Vice President or Director, 22% were production managers and R&D managers, 19% were sales managers, procurement managers and supply managers, and remaining 17% of respondents were other manager. So this composition reveals that most of respondents were knowledgeable about firm's supply chain management.

3.3 Missed Data

25% of received questionnaires included some missed data. So, an expectation maximization algorithm was used in Amelia II which is a recommended software for missed data imputation [33]. Prior to using expectation maximization, it must be assured that data were missing completely at random. Little's test for data in SPSS software, resulted in chai-sqare = 2385, df = 2428 and P = 0.725 which at confidence level of 0.05 means missing data were completely at random. So missed data were imputed with Amelia II and complete dataset for further analysis provided.

3.4 Reliability and Validity

In addition to content validity, mentioned in previous sections, the adequacy of a measure requires that three essential components be established: unidimensionality, reliability and validity [34]. Validity itself includes convergent validity and discriminant validity. So CFA was used for measurement model relevant tests. As the measurement model had more than four-point scales, based on [35] recommendation, the maximum likelihood method of LISREL was used for calculating model fit indexes, that is a more common and reliable method [35]. For assessing model fitting, two critical indexes of CFI and SRMR was used as recommended by [36] for less than 250 samples. The models were identified with CFI $\geq$ 0.95 and SRMR $\leq$ 0.09 as acceptable [36].

In the first stage, unidimensionality was tested, that involves establishment of a set of empirical indicators relates to one and only one construct [34]. A single factor LISREL measurement model was specified for all of constructs. If a construct had less than four items, two-factor model were tested by adding the items of another construct, making model fit indexes obtainable [31]. A CFA was conducted to separate measurement models of each construct, such as information sharing, strategic view in supply chain management and lean manufacturing practices. It was found that fitting indexes of some constructs were unsatisfactory. Then, the standardized residuals matrix of LISREL results were used to identify which items must be deleted to obtain better fit indexes for each model. Large standardized residuals indicate that a particular relationship is not well accounted by the model [37]. During this iterative procedure, one item out of measurement items of strategic view in supply chain management, lean manufacturing practices, performance management, general enablers, logistics and supply performance, and delivery performance were dropped. Also two items out of eight measurement items of integration were dropped. Table 3 shows the analysis results of the final structural model of all constructs.

In the second stage, the reliability analysis was conducted by using composite reliability (1) which is less sensitive to number of items of constructs [38].

$$\rho_\eta = \frac{\left(\sum_{i=1}^{p} \lambda_i\right)^2}{\left(\sum_{i=1}^{p} \lambda_i\right)^2 + \sum_{i=1}^{p} Var(\varepsilon_i)},\tag{1}$$

As depicted in Table 3, all of model constructs have an acceptable level of reliability, except production performance which its reliability index (ρ) is less than 0.7 cutoff criteria. SCP31 item was dropped from SC production performance construct to improve its reliability. So this construct finally reached the value of 0.9, which is a good level.

Table 3. Constructs properties for unidimesionality, reliability and convergent validity [4].

Constructs	χ^2	Df	CFI	SRMR	ρ	AVE
General SC enablers	57.70	26	0.97	0.05	0.84	0.65
Information sharing	22.24	8	0.95	0.06	0.78	0.73
Strategic view in supply chain management	6.47	5	0.99	0.03	0.76	0.62
Lean manufacturing practices	0.57	2	1.00	0.01	0.82	0.72
Supplier management	22.24	8	0.95	0.07	0.70	0.66
Performance management	7.43	2	0.96	0.05	0.70	0.59
SC Human resources management	33.45	8	0.96	0.04	0.72	0.75
Customer orientation	33.45	8	0.96	0.04	0.89	0.82
Supply chain integration	31.84	9	0.97	0.05	0.89	0.75
SC planning performance	41.12	10	0.96	0.04	0.90	0.95
SC logistics and supply performance	41.12	10	0.96	0.04	0.80	0.82
SC production performance	41.12	10	0.96	0.04	0.42	0.51
SC delivery performance	41.12	10	0.96	0.04	0.90	0.95
SC customer delight performance	41.12	10	0.96	0.04	0.86	0.89

In the third stage for analyzing construct validity, the convergent validity and discriminant validity were assessed. Convergent validity relates to the degree to which multiple methods of measuring a variable provide the same results [34]. Based on Fornell and Larcker [38] recommendation, the average variance extracted (AVE) was used to analyze convergent validity. An AVE greater than 0.5 is desirable because it suggests that on average, the latent construct accounts for a majority of the variance in its indicators [39]. Based on this criterion, as shown in Table 3 all research constructs have acceptable convergent validity.

For a measure to have discriminant validity, the variance in the measure should reflect only the variance attributable to its intended latent variable and not to other latent variables [34]. In analyzing discriminant validity for SC management practices, as recommended by Shiu et al. [40] both procedures of Fornell and Larcker [38], and Bagozzi and Phillips [41] were used. In doing first procedure, the squared correlation between a pair of constructs against the average variance extracted (AVE) for each of the two constructs was compared. For each pair of constructs, if the squared correlation was smaller than both the AVEs, it was concluded that the constructs exhibit discriminant validity. Based on the second procedure, the difference in chi-square value between the unconstrained CFA model and the nested CFA model was examined where the correlation between the target pair of constructs is constrained to unity. Based on these two procedures it was found out that all constructs have discriminant validity except the constructs of "Human resources management" and "Supplier management" which is one of limitations of this study.

3.5   Building Causal Bayesian Network

In this study Bayesian network was used. As stated by Heckerman [42], a Bayesian network can be used to learn causal relationships, and hence can be used to gain understanding about a problem domain and to predict the consequences of intervention. Furthermore, a Bayesian network model has both causal and probabilistic semantics, which is an ideal representation for combining prior knowledge and data.

To build a Bayesian network the data needs to be categorical. This way, the categorical measurements for each concept can be obtained by applying k-means cluster analysis [43]. In this study, Two-state categorization for the constructs of SC enabler and SCM practices, and three-state categorization for the constructs of SC performance were applied. For Bayesian causal modeling, TETRAD 6-4-0 is a program which creates, simulates data from, estimates, tests, predicts with, and searches for causal and statistical models [44] that is developed at Carnegie Mellon University.

In causal modeling process, first the categorical data was entered to TETRAD 6-4-0 package. Then, by using its knowledge module, the order of variables was specified. In Fig. 1, SC enablers are specified at first order and SCM practices at second and SC performance measures at last. In addition, it was specified that in each group of SC enablers and SCM practices, no inter-relationships be allowed by software, avoiding hyper-complex network.

4   Results

4.1   Causal Model

Running the PC algorithm with prior knowledge, as described in previous section, resulted in the model of Fig. 2. This model has degree of freedom of 152, chi-square of 624, and BIC of −180. In this primary model, production flexibility and customer satisfaction have no causal connection. It was suggested that some SC enablers may have direct impact on SC performance and some SC performance aspects may have effects on other SC performance aspects. Thus, the settings of the Search module of TETRAD 6-4-0 were modified for allowing the PC algorithm to find any direct relationships between SC enablers and SC performance aspects and also any relations between SC performance aspects. The resulted model (Fig. 3) has degree of freedom of 148, chi-square of 545 and BIC of −238.

At the first glance, it can be seen that advanced manufacturing technology such as SC enabler has direct impact on SC performance (delivery flexibility). In this model, delivery flexibility is antecedent of production flexibility and customer satisfaction. In addition, production flexibility is antecedent of logistics performance. This research suggests that the production flexibility must be antecedent of delivery performance, so this relation in resultant model was modified. The resultant model (Fig. 4) have degree of freedom of 148, chi-square of 546 and BIC of −236 which are totally better than previous model fit indices, verifying our modifications.

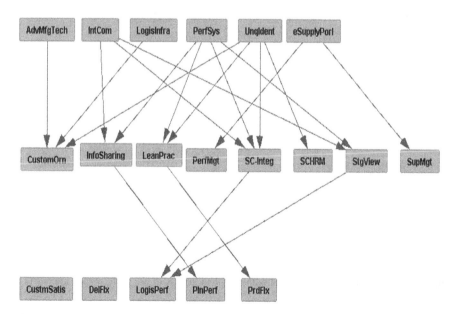

Fig. 2. Output of PC algorithm depicting causal Bayesian network of study variables [4].

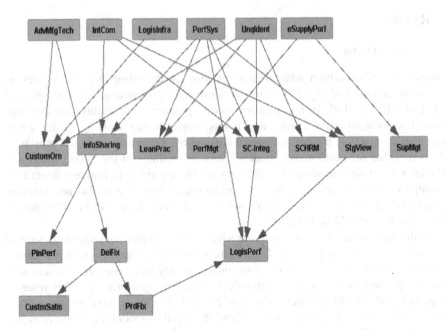

Fig. 3. Output of PC algorithm with modified prior knowledge [4].

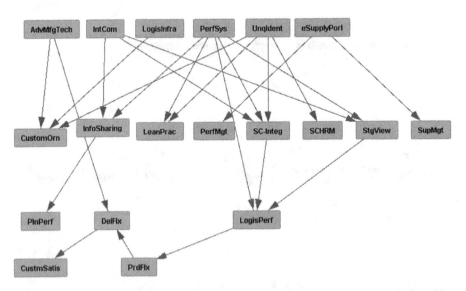

Fig. 4. Final bayesian network model with modified arrows of SC performance indices [4].

4.2   Bayesian Inference

For deepening the understanding of causal relations of the final model, conditional probability estimation with Bayesian networks, also known as Bayesian inference was developed. Probabilistic inference is concerned with revising probabilities for a variable or set of variables, called the query, when an intervention fixes the values of another variable or set of variables, called the evidence [45]. To do this job the maximum likelihood Bayes estimator module of TETRAD 6-4-0 software with its Dirichlet estimator was used to develop tables of conditional probabilities for SC enablers, SCM practices and SC performances of final CBN model. Dirichlet distribution is a generalization of beta distribution which is frequently used in Bayesian networks estimations.

Using the Dirichlet estimator, conditional tables for all of the model variables are developed. Figure 5, depicts the output of TETRAD 6-4-0 software for Dirichlet estimator which used for model variables. Some of the most important of them are presented and analyzed below.

Information Sharing. Information sharing is the first supply chain practice which its conditional table analyzed. As it can be seen in Table 4, information sharing as a SCM practice is conditional on performance management systems and inter-organizational communication technology as its enablers. Based on this table, when a supply chain has performances management systems and inter-organizational communication technology, it is more probable that an effective information sharing in that supply chain be available.

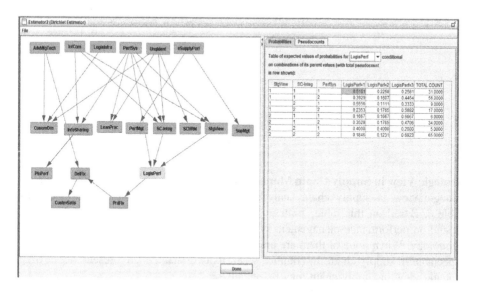

Fig. 5. Dirichlet estimator output of TETRAD 6-4-0 software.

Table 4. Conditional table of information sharing.

Performance management systems	Inter-organizational communication technology	Information sharing = 0	Information sharing = 1
0	0	0.7241	0.2759
0	1	0.6429	0.3571
1	0	0.5974	0.4026
1	1	0.4138	0.5862

Supply Chain Integration. Supply chain integration is one of the most discussed SCM practices [1, 7, 25, 46, 47]. As depicted in conditional Table 5, when its enablers are not present, there is a little chance for a supply chain to have effective supply chain integration. Also, when a supply chain has an inter-organizational communication but no effective performance management systems and unique identification and trace technologies are implemented, just 30% is probable that the supply chain integration be effective. But when all of the identified supply chain integration enablers are present, it can be expected that nearly 70% the supply chain integration be effective.

Table 5. Conditional table of supply chain integration.

Performance management systems	Inter-organizational communication technology	Unique identification and trace technologies	Supply chain integration = 0	Supply chain integration = 1
0	0	0	0.8077	0.1923
0	0	1	0.8000	0.2000
0	1	0	0.7000	0.3000
0	1	1	0.5000	0.5000
1	0	0	0.7069	0.2931
1	0	1	0.5238	0.4762
1	1	0	0.5405	0.4595
1	1	1	0.3077	0.6923

Strategic View in Supply Chain Management. As another important SCM practices, strategic view in supply chain analyzed, which its conditional table developed as Table 6. Based on this table, strategic view in supply chain management is strictly depend on performance management systems and inter-organizational communication technology. When none of them are present, just about 10% effective strategic view is expectable in supply chain. In contrast, when its two enablers are present, about 67% strategic view in supply chain may be effective.

Table 6. Conditional table of strategic view in supply chain management.

Performance management systems	Inter-organizational communication technology	Strategic view = 0	Strategic view = 1
0	0	0.8966	0.1034
0	1	0.7143	0.2857
1	0	0.5455	0.4545
1	1	0.3103	0.6897

Logistics Performance. Logistics performance is one of the most cited measures of supply chain performance. As shown in final causal model, this performance is also antecedent of other SC performance measures. This measure directly affected by performance management system as enabler, and supply chain integration and strategic view as SCM practices which related conditional probabilities are reported in Table 6. It must be noted that performance measure in this study have three levels including 0 as low level, 1 as mid-level and 2 as high or good level of performance.

At rows one to three in Table 7, strategic view isn't present so as expected, there are no good chance of high level performance of logistics. But at the forth row, when strategic view not present but the other antecedents are, there is about 0.59% chance for good logistics performance and in total 0.77 chance for acceptable logistics performance. At the last four rows of Table 6, it's clear that when strategic view in supply chain is present, the chance of good logistics performance is high conditional on presence of performance management systems (see row 7). It can be concluded that towards a better logistics performance, presence of the three antecedents increase the chance of good logistics performance to 69% and also its other affected performance measures. Also when strategic view and supply chain integration are present but no performance management systems, the chance of good logistics performance fall down to 20% which highlight the importance of performance management systems.

Table 7. Conditional table of logistics performance.

Strategic view	Supply chain integration	Performance management systems	Logistics performance = 0	Logistics performance = 1	Logistics performance = 2
0	0	0	0.5161	0.2258	0.2581
0	0	1	0.3929	0.1607	0.4464
0	1	0	0.5556	0.1111	0.3333
0	1	1	0.2353	0.1765	0.5882
1	0	0	0.1667	0.1667	0.6667
1	0	1	0.3529	0.1765	0.4706
1	1	0	0.4000	0.4000	0.2000
1	1	1	0.1846	0.1231	0.6923

5 Discussion and Implications

The resultant CBN model as discussed in work of Azhdari [4] in many aspects is supported by supply chain literature. This model (Fig. 4) show that advanced manufacturing technology and performance systems as SC enablers, and information sharing, SC integration and strategic view in supply chain as SCM practices have direct impact on SC performance measures such as logistics performance.

Using Bayesian inference to probabilistically analyzing the CBN relations revealed some interesting results. As it can be seen in Table 4, effective information sharing implementation needs both performance management systems and inter-organizational communication technologies in supply chain which the last was not considered before. In case of supply chain integration posterior knowledge inference it's found that when performance management systems are not effective or available in a supply chain, the chance of effective SC integration is just about 50%, despite of presence of inter-organizational communication and unique identification technologies, which clarify the importance of performance management systems. Also none of SC integration enablers by itself can significantly improve the chance of effective SC integration. The conditional table of strategic view (Table 6) discloses strategic view in supply chain can't be effective when its enablers including performance management systems and inter-organizational communication technologies like extranets are not implemented effectively.

Logistics performance is an important SC performance measure and also based on Fig. 4, sequentially has impact on some other SC performance measures. As presented in Table 7, when logistics performance antecedents including strategic view, supply chain integration and performance management systems are available, its chance of good performance is as high as about 70%. Also in total, it can be concluded that the most influential antecedent of logistics performance is strategic view in supply chain.

6 Conclusion and Limitations

In this research a causal model of supply chain enablers, practices and performance is developed and a Bayesian inference analysis used to deepen its results understandings. This work is a development of earlier work of Azhdari [4].

This study has some limitations regarding methodologies and scopes. First, the sample population was drawn from the members of the IranCode®. Although this sample covered a wide range of firms in terms of industry, size, and geography, it cannot be claimed that the results of this research can be wholly generalized, especially because the response rate was not high and this study were based on a self-assessment of the single participants from sample firms. So, further studies can be carried on for narrower group of industries with larger sample sizes. Because of a limited sample, some Bayesian inferences must be considered with caution. Causal sufficiency is a determinant in probabilistic causal modeling and therefore in Bayesian inference validity. Bayesian inference is based on conditional tables and when tables are more comprehensive, backward and forward inferences are more valid. Thus, it is needed to identify if any other contributing variables are neglected, which considering them may bring more valid causal models and related Bayesian inferences in this line of study.

In CBN model, some important intra-relations of SCM element's tier worth of further study, which ignoring them may blur the final results, especially weaken the Bayesian inferences. Particularly studying intra-relations between SCM practices may reveal many interesting results which contribute to more inclusive Bayesian inferences.

The set of SC performance measures were selected based on available data and some others eliminated because of measurement model validity. Hereafter, more definitive and comprehensive SC performance measurement may contribute to attaining more valid and applicable results from Bayesian inferences in the future studies.

Despite these limitations, this study has the following contributions in literature and practice. The first contribution of this study is its comprehensive review of supply chain enablers and supply chain management practices which as mentioned by [31], were not realized before. Second, as mentioned by Azhdari [4], a causal Bayesian network model is developed from field data and then using the TETRAD 6-4-0 tools, modified to better fit indices. Such a logical modification towards a better model fit indices is a new approach in methodology. At last, but the most important contribution of this study is applying Bayesian inference in SCM knowledge context. It is a new approach and its results contribute to deepening the knowledge of SCM dynamics and also make it more practical to SCM practitioners. As SCM practitioners can know in advance, which developments in SC enablers or SCM practices may result in which level of improvements in supply chain outcomes and to what extent? Also they can identify any SC performance weakness may due to which deficiencies in SC enablers or SCM practices or some combinations of them?

References

1. Hsu, C.C., Tan, K.C., Kannan, V.R., Keong Leong, G.: Supply chain management practices as a mediator of the relationship between operations capability and firm performance. Int. J. Prod. Res. **47**(3), 835–855 (2009)
2. Halley, A., Beaulieu, M.: Mastery of operational competencies in the context of supply chain management. Supply Chain Manage. Int. J. **14**(1), 49–63 (2009)
3. Li, S., Ragu-Nathan, B., Ragu-Nathan, T.S., Rao, S.S.: The impact of supply chain management practices on competitive advantage and organizational performance. Omega **34**, 107–124 (2006)
4. Azhdari, B.: Integrating fuzzy cognitive mapping and bayesian network learning for supply chain causal modeling. In: The 7th International Conference on Operations Research and Enterprise Systems, Funchal (2018)
5. Narasimhan, R., Jayanth, J.: Causal linkages in supply chain management: an exploratory study of North American manufacturing firms. Decis. Sci. **29**(3), 579–605 (1998)
6. Frohlich, M.T., Westbrook, R.: Arcs of integration: an international study of supply chain strategies. J. Oper. Manage. **19**, 185–200 (2001)
7. Tan, K.C., Lyman, S.B., Winser, J.D.: Supply chain management: a strategic perspective. Int. J. Oper. Prod. Manage. **22**(6), 614–631 (2002)

8. Li, S., Lin, B.: Accessing information sharing and information quality in supply chain management. Decis. Support Syst. **42**, 1641–1656 (2006)
9. Shin, H., Collier, D.A., Wilsom, D.D.: Supply management orientation and supplier/buyer performance. J. Oper. Manage. **18**, 317–333 (2000)
10. Lockamy III, A., McCormack, K.: Linking SCOR planning practices to supply chain performance: an exploratory study. Int. J. Oper. Prod. Manage. **24**(12), 1192–1218 (2004)
11. Lee, C.W., Kwon, I.-W.G., Severance, D.: Relationship between supply chain performance and degree of linkage among supplier, internal integration, and customer. Supply Chain Manage. Int. J. **12**(6), 444–452 (2007)
12. Sezen, B.: Relative effects of design, integration and information sharing on supply chain performance. Supply Chain Manage. Int. J. **13**(3), 233–240 (2008)
13. Ibrahim, S.E., Ogunyemi, O.: The effect of linkages and information sharing on supply chain and export performance: an empirical study of Egyptian textile manufacturers. J. Manufact. Technol. Manage. **23**(4), 441–463 (2012)
14. Li, G., Yang, H., Sun, L., Sohal, A.S.: The impact of IT implementation on supply chain integration and performance. Int. J. Prod. Econ. **120**, 125–138 (2009)
15. Zelbst, P.J., Green Jr., K.W., Swer, V.E., Baker, G.: RFID utilization and information sharing: the impact on supply chain performance. J. Bus. Ind. Mark. **25**(8), 582–589 (2010)
16. González-Benito, J.: Information technology investment and operational performance in purchasing: The mediating role of supply chain management practices and strategic integration of purchasing. Ind. Manage. Data Syst. **107**(2), 201–228 (2007)
17. Sanders, N.R.: An empirical study of the impact of e-business technologies on organizational collaboration and performance. J. Oper. Manage. **25**, 1332–1347 (2007)
18. Zhou, H., Benton Jr., W.C.: Supply chain practice and information sharing. J. Oper. Manage. **25**, 1348–1365 (2007)
19. Li, W., Humphreys, P.K., Yeung, A.C., Edwin Cheng, T.C.: The impact of specific supplier development efforts on buyer competitive advantage: an empirical model. Int. J. Prod. Econ. **106**, 230–247 (2007)
20. Johnson, P.F., Klassen, R.D., Leenders, M.R., Awaysheh, A.: Utilizing e-business technologies in supply chains: the impact of firm characteristics and teams. J. Oper. Manage. **25**, 1255–1274 (2007)
21. Devaraj, S., Krajewski, L., Wei, J.C.: Impact of eBusiness technologies on operational performance: the role of production information integration in the supply chain. J. Oper. Manage. **25**, 1199–1216 (2007)
22. Bayraktar, E., Demirbag, M., Koh, S.L., Tatoglu, E., Zaim, H.: A causal analysis of the impact of information systems and supply chain management practices on operational performance: evidence from manufacturing SMEs in Turkey. Int. J. Prod. Econ. **122**, 133–149 (2009)
23. Davis-Sramek, B., Germain, R., Karthik, I.: Supply chain technology: the role of environment in predicting performance. J. Acad. Mark. Sci. **38**, 42–55 (2010)
24. Sundram, V.P.K., Ibrahim, A.R., Govindaraju, V.C.: Supply chain management practices in the electronics industry in Malaysia: consequences for supply chain performance. Benchmarking: Int. J. **18**(6), 834–855 (2011)
25. Hamister, J.W.: Supply chain management practices in small retailers. Int. J. Retail Distrib. Manage. **40**(6), 427–450 (2012)
26. Ding, J., Krämer, B., Bai, Y., Chen, H.: Backward inference in bayesian networks for distributed systems management. J. Netw. Syst. Manage. **13**(4), 409–427 (2005)
27. Antai, I.: Supply chain vs supply chain competition: a niche-based approach. Manage. Res. Rev. **34**(10), 1107–1124 (2011)

28. Markis, S., Zoupas, P., Chryssolouris, G.: Supply chain control logic for enabling adaptability under uncertainty. Int. J. Prod. Res. **49**(1), 121–137 (2011)
29. Garvey, M.D., Carnovale, S., Yeniyurt, S.: An analytical framework for supply network risk propagation: a bayesian network approach. Eur. J. Oper. Res. **243**(2), 618–627 (2015)
30. Li, G., Yang, H., Sun, L., Sohal, A.S.: The impact of IT implementation on supply chain integration and performance. Int. J. Prod. Econ. **120**(1), 128–138 (2009)
31. Li, S., Rao, S.S., Ragu-Nathan, T.S., Ragu-Nathan, B.: Development and validation of a measurement instrument for studying supply chain management practices. J. Oper. Manage. **23**, 618–641 (2005)
32. Bagozzi, R.P.: Measurement and meaning in information systems and oraganizational research: methodological and philosophical foundations. MIS Q. **35**(2), 261–292 (2011)
33. King, G., Honaker, J., Joseph, A., Scheve, K.: Analyzing incomplete political science data: an alternative algorithm for multiple imputation. Am. Polit. Sci. Rev. **95**(1), 49–69 (2001)
34. O'Leary-Kelly, S.W., Vokurka, R.J.: The empirical assessment of construct validity. J. Oper. Manage. **16**, 387–405 (1998)
35. Bentler, P.M., Chou, C.-P.: Practical issues in structural modeling. Sociol. Methods Res. **16**(1), 78–117 (1987)
36. Hu, L.-T., Bentler, P.M.: Cutoff criteria for fit indexes in covariance structure analysis: conventional criteria versus new alternatives. Struct. Eqn. Model. Multi. J. **6**(1), 1–55 (1999)
37. Schumacker, R.E., Lomax, R.G.: A Beginner's Guide to Structural Equation Modeling, 2nd edn. Lawrence Erlbaum Associates, Mahwah (2004)
38. Fornell, C., Larcker, D.F.: Evaluating structural equation models with unobservable variables and measurement error. J. Market. Res. **18**(1), 39–50 (1981)
39. MacKenzie, S.B., Podsakoff, P.M.: Construct measurement and validation procedures in MIS and behavioral research: integrating new and existing techniques. MIS Q. **35**(2), 293–334 (2011)
40. Shiu, E., Pervan, S.J., Bove, L.L., Beatty, S.E.: Reflections on discriminant validity: reexamining the Bove et al. (2009) findings. J. Bus. Res. **64**, 497–500 (2011)
41. Bagozzi, R.P., Phillips, L.: Representing and testing organizational theories: a holistic construal. Adm. Sci. Q. **27**, 459–489 (1982)
42. Heckerman, D.: Bayasian networks for data mining. Data Min. Knowl. Disc. **1**, 79–119 (1997)
43. McColl-Kennedy, J.R., Anderson, R.D.: Subordinate-manager gender combination and perceived leadership style influence on emotions, self-esteem and organizational commitment. J. Bus. Res. **58**, 115–125 (2005)
44. The Tetrad Project. http://www.phil.cmu.edu/tetrad/. Accessed 03 May 2018
45. Anderson, R.D., Vastag, G.: Causal modeling alternatives in operations research: overview and application. Eur. J. Oper. Res. **156**, 92–109 (2004)
46. Attaran, M.: RFID: an enabler of supply chain operations. Supply Chain Manage. Int. J. **12**(4), 249–257 (2007)
47. Banomyong, R., Supatn, N.: Developing a supply chain performance tool for SMEs in Thailand. Supply Chain Manage. Int. J. **16**(1), 20–31 (2011)

Operations Administration and Maintenance Constraints in Fiber Cables Network Design

Vincent Angilella[1,2]($\boxtimes$), Matthieu Chardy[1], and Walid Ben-Ameur[2]

[1] Orange Labs, 44 avenue de la République, 92320 Chatillon, France
vincent1.angilella@gmail.com
[2] SAMOVAR, Télécom SudParis, CNRS, Université Paris-Saclay,
9 rue Charles Fourier, 91011 Evry Cedex, France

Abstract. We introduce two specific design problems of optical fiber cable networks that differ by a practical maintenance constraint. An integer programming based method including valid inequalities is introduced for the unconstrained problem. We propose two exact solution methods to tackle the constrained problem: the first one is based on mixed integer programming including valid inequalities while the second one is built on dynamic programming. We then provide a fully polynomial time approximation scheme for the constrained problem. The theoretical complexities of both problems in several cases are proven and compared. Numerical results assess the efficiency of both methods in different contexts including real-life instances, and evaluate the effect of the maintenance constraint on the solution quality.

Keywords: Optical networks · Network design ·
Mixed integer programming · Dynamic programming

1 Introduction

Fiber To The Home (FTTH) networks are currently deployed by telecommunications operators, and require a huge capital expenditure (see [7], it can cost several billion euros to connect one million households). The technological architecture chosen by a majority of operators is to deploy passive optical networks, which are based on passive optical splitters. A passive optical splitter connects several fibers on one of its sides to one at the other side (divides or gathers the signal depending on its origin), which leads to a tree topology of the FTTH networks (illustrated in Fig. 1a). The design of such networks includes to decide the splitter locations, the civil engineering infrastructure used (see [4–6,8]). Finally, the fiber cable network has to be designed to connect these equipment (see Fig. 1a). These decisions are usually taken in different steps.

This paper focuses on the problem of fiber cable network design. This problem is highlighted in the survey [9] as an incomplete field of study, especially

© Springer Nature Switzerland AG 2019
G. H. Parlier et al. (Eds.): ICORES 2018, CCIS 966, pp. 54–79, 2019.
https://doi.org/10.1007/978-3-030-16035-7_4

when cable separation techniques are considered. The work from [1] tackles the issue including the selection of civil engineering infrastructure, but faces computational limits on real-life instances. The paper [12] excludes weld costs, which are a significant expense source. The work from [2] deals with the issue of cable backfeed, specific to the problem, but restricts the possible ways to serve the demand. In the following we include several ways to serve the demand (with fiber cables or fiber modules), and introduce a maintenance constraint which, to our knowledge, is novel. What follows extends the work presented in [3].

The next section introduces two problems which differ by the introduction of an Operation Administration & Maintenance constraint. We introduce an algorithm based on integer programming for the unconstrained problem in Sect. 3.1. Two solution methods are then proposed for the constrained problem, an integer programming based solution in Sect. 3.2, and a dynamic programming based solution in Sect. 4.1. A fixed parameter tractable approximation scheme is introduced for the constrained problem in Sect. 4.2. The theoretical complexities of both problems are proven and argued in Sect. 5. All solution methods are assessed numerically in Sect. 6.

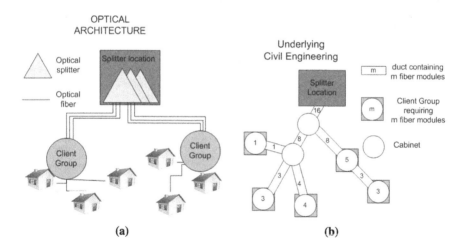

Fig. 1. (a) Underlying optical architecture example. It has a tree topology; the splitter location is connected to every client group [3]. (b) Underlying civil engineering tree example. The ducts, cabinets, demands and number of fiber modules are known [3].

2   Problem Description

The general problem tackled in this paper consists in connecting one splitter location to several client groups, using fiber cables, with minimal cost. It arises several times in a given FTTH network, notably once for each splitter location.

2.1 Unconstrained Problem

Cables are to be laid out in a civil engineering infrastructure (usually the one used for the legacy copper network) with a tree topology, assumed chosen within previous decision steps. The cables have an arborescent structure from the splitter location to the client groups. Along the ducts of this infrastructure are located street cabinets, in which the demand lies. The civil engineering structure used is supposed to be known due to previous decision making, as well as the demand in each cabinet.

Fiber cables contain several fiber modules, and each fiber module contains several fibers. Due to operational constraints, modules are not dividable, and all modules on a given network are supposed to be identical. This allows us to consider only fiber modules, and ignore the fiber level. Some of the modules are connected to the fiber source on one of their ends, and on the fiber demand on the other end. These are actually used, and are called "active modules", the other ones are called "dead modules". The latter can arise due to cables not matching exactly the demand or in the operations described below (example: a 4 module cable serving a cabinet which requires 3 modules). Since all the demand is known and there is only one path from the source to a given demand point, the number of active modules that must be deployed through a given duct is known (see Fig. 1b).

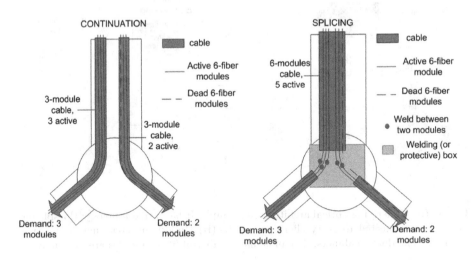

Fig. 2. Left: Continued cables; Right: Splicing operation [3].

At a cabinet, cables can endure a splicing operation, which leads to two basic configurations (see Fig. 2):

- All cables are continued. One only has to pay for the cost of laying out cables.
- One cable is spliced. It is cut at the cabinet, and its active modules are welded to active modules of new cables, referred to as "born cables". A protective

box, the size of which depends on the spliced cable size, is installed. One has to pay for the cables, the box and the welds.

There are two different ways to serve the demand that cannot be combined (see Fig. 3):

- Cable-served. In this case, a single cable brings all the required active modules to the demand cabinet.
- Module-served. In this case, a splicing operation is done in the cabinet, and some modules from the spliced cable are used to serve the demand. No welds are done on these modules.

Additional engineering rules have to be taken into account:

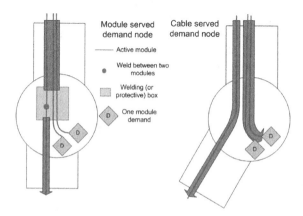

Fig. 3. Left: Module-served demand node; Right: Cable-served demand node [3].

- At most one cable can be spliced at a street cabinet. This is due to space restrictions and regulatory constraints (protective boxes are large).
- The demand of a given cabinet must be served by at most one cable.

The cost elements are as follows:

- The cost of a cable is linear with respect to its length, and concave with respect to its size (i.e. its number of modules). This derives from the catalogues of cable manufacturers, who propose a fixed price per length unit for each cable size.
- The cost of a protective box depends on the size of the cable being spliced. It is a piecewise constant function. This derives from the number of different boxes sold by manufacturers.

– The cost of welds depends on the number of welds to be done in a given
cabinet. It is piecewise linear concave, and derives from manpower cost considerations.

This decision problem, referred to as FCNDA (Fiber Cable Network Design
in an Arborescence) in the following, can be formulated as follows: given a civil
engineering arborescence, demand nodes, a set of available cables and the associated costs, design a minimum cost optical fiber cable network satisfying the
engineering rules listed above.

Section 2.2 introduces a restriction of the FCNDA problem.

2.2   Constrained Problem

We restrict the problem by imposing that all cables going through a given duct
are born in the same cabinet (eventually the fiber source). This restriction is
illustrated in Fig. 4. It is motivated by operations and maintenance considerations. Indeed, assuming all the cables of a given duct are damaged, then an
intervention has to be done at the cabinets where each of these cables is born.
If the rule is respected, an intervention is necessary in only one cabinet.

The constrained decision problem, referred to as EFCNDA (Easy-
maintenance Fiber Cable Network Design in an Arborescence) in the following
consists in designing a FCNDA solution where cables on a same duct are born
in the same cabinet with minimal cost.

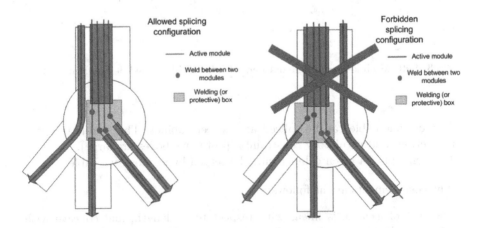

Fig. 4. Left: Allowed splicing configuration for EFCNDA. On all edges, cables are
born in the same cabinet; Right: Forbidden splicing configuration for EFCNDA. On
the bottom-right duct, two different cables are born in different cabinets [3].

3 Integer Programming

3.1 SFCND

Notation and Formulation. The following notations will also be used in Sect. 4.

An arborescence $G = (V, A)$ describes the civil engineering infrastructure, V the cabinets and A the ducts, and its root $r \in V$ denotes the fiber source (CO or splitter location). For any $i \in V$, $D_i \in \mathbb{N}$ denotes the demand (number of active modules required) in node i. We define $V^* = V \setminus r$, the set of demand nodes is noted $V_D = \{v \in V, D_v > 0\}$, the set of nodes without demand $V_N = V^* \setminus V_D$. Each arc $(i, j) \in A$ has a length $\Delta_{(i,j)} > 0$ and must contain $m_{i,j}^{act}$ active modules ($m_{i,j}^{act}$ being known, since we are in an arborescence). For $i \in V$, we denote $\Gamma^+(i)$ the set successors of i and $\gamma(i)$ its predecessor.

We have L different cable types at our disposal, we note $\mathcal{L} = \{1, .., L\}$ the set of cables. Cables of type $l \in \mathcal{L}$ have a size of $M_l \in \mathbb{N}$ modules, and for $l \in \mathcal{L}$, we note $\mathcal{M}_l = \{1, .., M_l\}$ (the range of possible number of active modules in a cable of type l).

For $l \in \mathcal{L}$, let us define C_l^{le} the cost per length unit of a cable of size l, and PB_l the cost of a box of size l. For $m \in \mathcal{M}_L$, let us define the cost of the smallest cable able to contain m active modules $C_m^{min} = C_{l_1}^{le}$ where $l_1 = \min\{l \in \mathcal{L}, m \le M_l\}$, and PW_m the cost for welding m modules.

We introduce $\mathcal{P}$ the set of directed paths of G, and for $p \in \mathcal{P}$, we note by $s(p)$ its source node, $t(p)$ its target node, and Δ_p its length (which extends Δ from A to $\mathcal{P}$).

We define the following variables:

- $\forall l \in \mathcal{L}, \forall p \in \mathcal{P}, k_{p,l}^{spl} \in \{0, 1\}$ the binary variable equal to 1 iff there is a cable of size l on path p spliced in $t(p)$.
- $\forall p \in \mathcal{P}, k_p^{dem} \in \{0, 1\}$ the binary variable equal to 1 iff there is a cable on path p serving the demand in $t(p)$ in a cable-served way. Its size is known, it is $\min\{l \in \mathcal{L} | M_l \ge D_{t(p)}\}$.
- $\forall p \in \mathcal{P}, m_p^{spl} \in \{0, .., M_L\}$ the number of active modules of the cable on path p spliced in $t(p)$.
- $\forall i \in V^*, \forall m \in \mathcal{M}_L, w_{i,m}$ the binary variable equal to 1 iff m welds are done in node i.

The problem can be formulated as follows:

$$\min \sum_{p \in \mathcal{P}} \Delta_p \cdot \left(C_{D_{t(p)}}^{min} \cdot k_p^{dem} + \sum_{l \in \mathcal{L}} C_l^{le} \cdot k_{p,l}^{spl} \right)$$
$$+ \sum_{i \in V_N} \sum_{m \in \mathcal{M}_L} PW_m \cdot w_{i,m} + \sum_{p \in \mathcal{P}} \sum_{l \in \mathcal{L}} PB_l \cdot k_{p,l}^{spl}$$

such that

$$\sum_{p\in\mathcal{P}|t(p)=i}\sum_{l\in\mathcal{L}}k^{spl}_{p,l} \leq 1 \qquad\qquad \forall i \in V^*, \qquad (1)$$

$$\sum_{p\in\mathcal{P}|t(p)=i}k^{dem}_p \leq 1 \qquad\qquad \forall i \in V_D, \qquad (2)$$

$$\sum_{l\in\mathcal{L}}M_l \cdot k^{spl}_{p,l} \geq m^{spl}_p \qquad\qquad \forall p \in \mathcal{P}, \qquad (3)$$

$$\sum_{p\in\mathcal{P}|t(p)=i}m^{spl}_p = D_i \cdot (1 - \sum_{p\in\mathcal{P}|t(p)=i}k^{dem}_p)$$

$$+ \sum_{p\in\mathcal{P}|s(p)=i}(m^{spl}_p + D_{t(p)}k^{dem}_p) \qquad \forall i \in V^*, \qquad (4)$$

$$\sum_{m\in\mathcal{M}_L}m \cdot w_{i,m} = \sum_{p\in\mathcal{P}|i=s(p)}(m^{spl}_p + D_{t(p)} \cdot k^{dem}_p) \qquad \forall i \in V^*, \qquad (5)$$

$$\sum_{m\in\mathcal{M}_L}w_{i,m} \leq 1 \qquad\qquad \forall i \in V_N,$$

$$k^{dem}, k^{spl}, w \in \{0,1\}; m^{spl} \in \{0,..,M_L\} \qquad\qquad (6)$$

In the cost function, the first term stands for the cost of cables, the second term for the cost of welds, and the last term for the cost of boxes. Equations (1) ensure at most one cable is spliced in a node. Constraints (2) a most one cable serves the demand in a cable-served way. Equations (3) make sure that spliced cables are large enough to contain their number of active modules. Constraints (4) are active module conservation equations. The left hand side term stands for the number of modules of the spliced cable. The first right side hand term is the number of modules necessary to serve the demand, in case it is not cable-served. The last term is the number of active modules of born cables. Finally, (5) and (6) ensure that w counts the number of welds to be done in each node.

Remark 1. It is possible to fix the value of some variables. First, notice that leaf nodes are demand nodes. These nodes will be served in a cable-served way, and no operation will be done inside them. This gives, for all nodes $i \in V_D$ such that $|\Gamma^+(i)| = 0$:

$$\forall m \in \mathcal{M}_L, w_{i,m} = 0$$

$$\forall p \in \mathcal{P}|t(p) = i, \forall l \in \mathcal{L}, k^{spl}_{p,l} = 0$$

Furthermore, the number of welds done in a node cannot exceed the number of active modules going out of this node. This gives:

$$\forall i \in V^*, \forall m \in \mathcal{M}_L, \text{ if } m > \sum_{j \in \Gamma^+(i)} m_{i,j}^{act}, w_{i,m} = 0$$

Valid Inequalities. We propose here several valid inequalities to tighten the formulation.

Let us define, for all $m \in \mathbb{N}$, the minimum cost per length unit of a set of cables able to contain m active modules denoted by $LB(m)$. For a given m, $LB(m) = \{\min \sum_{l \in \mathcal{L}} C_l^{le} \cdot n_l | \sum_{l \in \mathcal{L}} M_l \cdot n_l \geq m, n \in \mathbb{N}^L\}$.

Proposition 1. *The following inequalities are valid for the ESFCND problem:*

$$\forall (i,j) \in A, \quad \sum_{p \in \mathcal{P}|(i,j) \in p} \left(\sum_{l \in \mathcal{L}} (C_l^{le} \cdot k_{p,l}^{spl}) \right.$$
$$\left. + C^{min}(D_{t(p)}) \cdot k_p^{dem} \right) \geq LB(m_{i,j}^{act}) \qquad (7)$$

The left hand side is the cost per length unit of the cables going through (i, j).

Let us consider a path $p \in \mathcal{P}$ such that $t(p) \in V_D$ and $s(p) \neq r$. If there is a cable deployed on p, born in $s(p)$ and serving the demand in $t(p)$, then we know there is a splicing operation done in $s(p)$. Furthermore, there is at least $D_{t(p)}$ welds in this operation, since the cable serving $t(p)$ contains $D_{t(p)}$ active modules.

Proposition 2. *The following valid inequalities are valid for the ESFCND problem:*

$$\forall p \in \mathcal{P}|t(p) \in V_D \text{ and } s(p) \neq r, k_p^{dem} \leq \sum_{m \geq D_{t(p)}} w_{s(p),m} \qquad (8)$$

Proof. Let consider a path $p \in \mathcal{P}$ such that $t(p) \in V_D$, $s(p) \neq r$, and $k_p^{dem} = 1$ ($t(p)$ is cable served by a cable on p). By (5), it gives $\sum_{m \in \mathcal{M}_L} m \cdot w_{s_p,m} \geq D_{t(p)}$ (there are at least $D_{t(p)}$ welds done in $s(p)$). Which means, with (6), $\exists! m_0 \geq D_{t(p)}, w_{s_p,m_0} = 1$. Hence the result.   □

3.2 ESFCND

ESFCND can be solved by using the same variables as in Sect. 3.1. The cost function is the same, the set of feasible solutions is described by constraints (1) to (6) to which we add the maintenance constraints described below:

$$\forall (p, p') \in \mathcal{P}^2 \text{ such that } s(p) \neq s(p') \text{ and } \exists a \in A, a \in p \text{ and } a \in p',$$

$$k_p^{dem} + k_{p'}^{dem} \leq 1 \tag{9}$$

$$\sum_{l \in \mathcal{L}} k_{p,l}^{spl} + \sum_{l \in \mathcal{L}} k_{p',l}^{spl} \leq 1 \tag{10}$$

$$\sum_{l \in \mathcal{L}} k_{p,l}^{spl} + k_{p'}^{dem} \leq 1 \tag{11}$$

These constraints ensure that on two paths which have different origins but an arc in common, there can be only one cable. Constraints (9) ensure it in the case the two cables are serving the demand. Constraints (10) in the case both cables are spliced (at most one term in the sum $\sum_{l \in \mathcal{L}} k_{p,l}^{spl}$ is equal to 1, since there can be at most one splicing operation in $t(p)$, the same goes for p'). Finally, constraints (11) in the case one of them is spliced and the other one serves the demand.

The next section introduces an alternative mixed integer programming approach for ESFCND, based on arcs rather than paths. It uses the properties of the problem, and has less variables and less constraints.

Notations and Formulation. We keep the same notations for the problem instance. In addition, let us define for $(i, j) \in A, U_{i,j}$ an upper bound of the cost per length unit of the cables going through duct (i, j).

We define the following variables:

- $\forall (i, j) \in A, x_{i,j} \in \{0, 1\}$ the binary variable equal to 1 iff the cables on arc (i, j) are born in i.
- $\forall (i, j) \in A, c_{i,j} \in \mathbb{R}$ the continuous variable equal to the cost per length unit of the cables on arc (i, j).
- $\forall (i, j) \in A, z_{i,j} \in \mathbb{R}$ the continuous variable equal to $x_{i,j} \cdot c_{i,j}$.
- $\forall i \in V_D, u_i \in \{0, 1\}$ the binary variable equal to 1 iff the node i is module-served.
- $\forall i \in V^*, \forall m \in \mathcal{M}_L, w_{i,m}$ the binary variable equal to 1 iff m welds are done in node i (since its meaning is identical to Sect. 3.1, we keep the same name).
- $\forall i \in V^*, \forall l \in \mathcal{L}, y_{i,l}$ the binary variable equal to 1 iff a cable of size l is spliced in i.

The problem can be formulated as follows:

$$\min \sum_{i \in V^*} \sum_{m \in \mathcal{M}_L} PW_m \cdot w_{i,m}$$

$$+ \sum_{(i,j) \in A} \Delta_{(i,j)} \cdot c_{i,j} + \sum_{i \in V^*} \sum_{l \in \mathcal{L}} PB_l \cdot y_{i,l} \tag{12}$$

such that

$$c_{\gamma(i),i} = \sum_{l \in \mathcal{L}} C_l^{le} y_{i,l} + \sum_{j \in \Gamma^+(i)} c_{i,j}$$
$$- \sum_{j \in \Gamma^+(i)} z_{i,j} + (1 - u_i) \cdot C_{D_i}^{min} \qquad \forall i \in V_D, \qquad (13)$$

$$c_{\gamma(i),i} = \sum_{l \in \mathcal{L}} C_l^{le} y_{i,l} + \sum_{j \in \Gamma^+(i)} c_{i,j} - \sum_{j \in \Gamma^+(i)} z_{i,j} \qquad \forall i \in V_N, \qquad (14)$$

$$\sum_{l \in \mathcal{L}} M_l \cdot y_{i,l} \geq D_i \cdot u_i + \sum_{j \in \Gamma^+(i)} m_{i,j}^{act} \cdot x_{i,j} \qquad \forall i \in V_D, \qquad (15)$$

$$\sum_{l \in \mathcal{L}} M_l \cdot y_{i,l} \geq \sum_{j \in \Gamma^+(i)} m_{i,j}^{act} \cdot x_{i,j} \qquad \forall i \in V_N, \qquad (16)$$

$$\sum_{l \in \mathcal{L}} y_{i,l} \leq 1 \qquad \forall i \in V^*, \qquad (17)$$

$$\sum_{m \in \mathcal{M}_L} m \cdot w_{i,m} = \sum_{j \in \Gamma^+(i)} m_{i,j}^{act} \cdot x_{i,j} \qquad \forall i \in V^*, \qquad (18)$$

$$\sum_{m \in \mathcal{M}_L} w_{i,m} \leq 1 \qquad \forall i \in V^*, \qquad (19)$$

$$z_{i,j} \geq c_{i,j} - U_{i,j} \cdot (1 - x_{i,j}) \qquad \forall (i,j) \in A, \qquad (20)$$
$$z_{i,j} \leq U_{i,j} \cdot x_{i,j} \qquad \forall (i,j) \in A, \qquad (21)$$
$$z_{i,j} \leq c_{i,j} \qquad \forall (i,j) \in A, \qquad (22)$$
$$u, w, x, y \in \{0,1\}; c, z \in \mathbb{R}$$

The first term of the cost function denotes the cost of welds, the second term stands for the cost of cables, and the last term stands for the cost of boxes. Equations (13) ensure the cost per length unit of any arc is properly counted. The term $\sum_{l \in \mathcal{L}} C_l^{le} y_{i,l}$ stands for the cost of the cable spliced in i, if any. If for some arc $(i,j) \in A$ such that $j \in \Gamma^+(i)$ we have $x_{i,j} = 0$, then the cables on (i,j) come from $(\gamma(i), i)$ unchanged. Otherwise, they come from the splicing operation done in i. The last term stands for the cost of the cable serving the demand in i. Equations (14) are the equivalent concerning nodes without demand. Equations (15), (16) and (17) ensure the cable spliced in i is large enough to contain its active modules. The first term of the right hand side of (15) stands for modules serving the demand, the second term for modules of born cables. Constraints (18) and (19) ensure the variable $w_{i,m}$ is equal to 1 iff there are m welds done in node i. Finally, constraints (20), (21) and (22) ensure $\forall (i,j) \in A, z_{i,j} = x_{i,j} \cdot c_{i,j}$ (these are linearisation equations).

Remark 2. It is possible to fix the value of some variables. Assuming there exists $i \in V^*$ and $m_1 \in \mathcal{M}_L$ such that $w_{i,m_1} = 1$, then by (18), we know there exists $S \subseteq \Gamma^+(i)$ such that $m_1 = \sum_{j \in S} m_{i,j}^{act}$. This gives by contraposition $\forall i \in V^*$, $\forall m \in \mathcal{M}_L$ if $m \notin \{\sum_{j \in S} m_{i,j}^{act} | S \subseteq \Gamma^+(i)\}$ then $w_{i,m} = 0$. It can be computed in $\mathcal{O}(|\Gamma^+(i)| \times M_L)$ (which is not a polynomial with respect to the instance size, provided M_L is not coded in an unary system).

Valid Inequalities. The continuous relaxation of the formulation introduced above shows is weak, mostly due to the linearisation of z. We propose here several valid inequalities to tighten it.

In nodes without demand, if a cable of size l is spliced, then it has a number of active modules between M_l and $M_{l-1}+1$; otherwise one could install a smaller cable and obtain a cheaper solution. With the convention $M_0 = 0$ and $\mathcal{M}_0 = \emptyset$, this gives:

Proposition 3. *Every optimal solution of the ESFCND problem verifies*

$$\forall i \in V_N, \forall l \in \mathcal{L}, y_{i,l} = \sum_{m \in \mathcal{M}_l \setminus \mathcal{M}_{l-1}} w_{i,m} \tag{23}$$

Proof. Let us consider an optimal solution S of the ESFCND problem. Let us consider $i \in V^*$ and $l \in \mathcal{L}$ such that $y_{i,l} = 1$ (a box of size l is installed in i). This gives us $1 \le \sum_{j \in \Gamma^+(i)} m_{i,j}^{act} \cdot x_{i,j}$ (there are cables born in i); otherwise we could obtain a cheaper solution by setting $y_{i,l}$ to 0.

Either (16) or (15) give us $M_l \ge \sum_{j \in \Gamma^+(i)} m_{i,j}^{act} \cdot x_{i,j}$. Furthermore, with (18) and (19), we can obtain $\exists m_0 \in \{1, .., M_l\}, w_{i,m_0} = 1$ (in other words, $m_0 \le M_l$ welds are done in i).

If $l = 1$, we have the result.

Otherwise, let us assume $m_0 \le M_{l-1}$. Then, the solution S' identical to S everywhere but in $y'_{i,l-1} = 1$ and $y'_{i,l} = 0$ is a feasible cheaper solution (it is the solution obtained by replacing the cable spliced in i by a smaller cable, leading to a smaller cost for boxes and cables). Which contradicts our hypothesis.

Hence the result.   □

With a reasonment similar to the one from Proposition 1 (see definition of LB), we can get a lower bound of the cost per length unit of the cables on each arc.

Proposition 4. *The following inequalities are valid for the ESFCND problem:*

$$\forall (i,j) \in A, c_{i,j} \ge LB(m_{i,j}^{act}) \tag{24}$$

If the cables on some arc $(i,j) \in A$ are born in i, then at least $m_{i,j}^{act}$ welds are done in node i. This implies what follows.

Proposition 5. *The following inequalities are valid for the ESFCND problem*

$$\forall (i,j) \in A, x_{i,j} \le \sum_{m \in \mathcal{M}_L | m \ge m_{i,j}^{act}} w_{i,m} \tag{25}$$

Proof. Let us consider a solution of the ESFCND problem. Let us consider $(i,j) \in A$ such that $x_{i,j} = 1$. This implies, by (18) that $\sum_{m \in \mathcal{M}_L} m \cdot w_{i,m} \ge m_{i,j}^{act}$. Then, with (19), it follows that $\exists! m_0 \ge m_{i,j}^{act}, w_{i,m} = 1$ (only one of the variables $w_{i,m}$ can be equal to 1). Hence the result.   □

4 Dynamic Programming for ESFCND

For any node $i \in V^*$, we introduce the additional notation $V^{pr}(i)$, which refers to the set of nodes on the path from the root to i, excluding i and including r.

4.1 Exact Algorithm

The ESFCND problem can be solved by Algorithm 1. To each node $i \in V^*$, and for each node $j \in V^{pr}(i)$, we associate to i a label $< \ j, C(i,j) > \in V^{pr}(i) \times \mathbb{R}$ where $C(i,j)$ is the minimum cost of the network rooted in i plus the cost of the cables on the path from j to i, assuming these are born in node j.

Algorithm 1. Exact Resolution Algorithm for ESFCND.

1: **procedure** INITIALISATION()
2:     **for** $i \in V_D | \Gamma^+(i) = \emptyset$ **do**
3:         **for** $j \in V^{pr}(i)$ **do**
4:             Add to i the label $< j, C_{D_i}^{min} \cdot \Delta_p >$ where $p \in \mathcal{P}$ is the only path s.t. $s(p) = j$ and $t(p) = i$.
5:         **end for**
6:         Declare i labeled.
7:     **end for**
8: **end procedure**
9: **procedure** RECURSION()
10:     **while** $\exists r' \in \Gamma^+(r)$ such that r' has not been labeled **do**
11:         **for** every node $i \in V^*$ such that all nodes in $\Gamma^+(i)$ have been labeled **do**
12:             **for** $j \in V^{pr}(i)$ **do**
13:                             ▷ We select the operation in i minimizing the network cost.
14:                 Add the label $< j, C(i,j) >$ to node i where

$$C(i,j) = \min_{S \subseteq \Gamma^+(i), u \in \{0,1\}} \sum_{k \in S} C(k,i) + \sum_{k \in \Gamma^+(i) \setminus S} C(k,j)$$
$$+ PW_m + \Delta_p \cdot C_{l_1}^{le} + \Delta_p \cdot C_{D_i}^{min} \cdot (1 - u) \qquad (26)$$

$$\text{with} \quad \begin{cases} m = \sum_{k \in S} m_{i,k}^{act}; l_1 = \min\{l \in \mathcal{L} | M_l \geq u \cdot D_i + \sum_{k \in S} m_{i,k}^{act}\} \\ p \in \mathcal{P} \text{ is the only path such that } s(p) = j, t(p) = i \end{cases}$$

15:             **end for**
16:             Declare i labeled.
17:         **end for**
18:     **end while**
19: **end procedure**
20: **procedure** TERMINATION()
21:     **return** $\sum_{r' \in \Gamma^+(r)} C(r', r)$
22: **end procedure**

The algorithm is initialized at leaf nodes (line 4), which are cable-served demand nodes, and where the size of the cable serving the demand is known.

For a node i such that all nodes in $\Gamma^+(i)$ have been labeled, and for $j \in V^{pr}(i)$, (26) computes the minimum cost network if the next operation is done in j. For $i \in V^*$ and $k \in \Gamma^+(i)$, $k \in S$ iff the cables going through arc (i, k) are born in node i. Similarly, the boolean u is equal to 1 iff the node i is module-served (its meaning is similar than the variable u_i in Sect. 3.2).

We propose to compute it with a brute-search algorithm on the set S and on u. For given nodes $i \in V^*, j \in V^{pr}(i)$, it can be done in $\mathcal{O}(|\Gamma^+(i)| \times 2^{|\Gamma^+(i)|+1})$.

Lemma 1. *Algorithm 1 runs in time $\mathcal{O}(2^{1+\max \Gamma} \times |V|^2)$ where $\max \Gamma$ denotes the maximal degree (number of successors) of a node in the graph.*

This can be shown by summing the operations done for each loop.

Remark 3. This implies that if the maximal degree of nodes in the graph is bounded by a constant, then Algorithm 1 runs in polynomial time.

For a non-leaf node $i \in V^*$ and $j \in V^{pr}(i)$, when we compute (26), we do not consider the cost of the welds done in j. This comes later, while j is being labeled. It does not influence the network below, since all cables going through $(\gamma(i), i)$ are born in i. C^* is the sum of the following elements:

- the cost of the network in the arborescence rooted in i, including the cost of the welds and boxes in i (if any)
- the cost of cables deployed from i to j

This leads us to show the next proposition to show the validity of the algorithm.

Proposition 6. *Let us consider $i \in V^*$. When i is declared labeled in Algorithm 1, there exists a node $j \in V^{pr}(i)$ such that in the label $< j, C(i, j) >$, $C(i, j)$ describes the cost of the minimum ESFCND solution in the arborescence rooted in node i plus the cost of the cables on the path from j to i.*

We will start to prove it for leaf nodes, then recursively on higher nodes.

Proof. ⋆ Let us consider a leaf node i. In the minimum cost network, it is served in a cable-served way with a cable of type $l_1 = \min\{l \in \mathcal{L} | M_l \geq D_i\}$. This cable is born in some node $j \in V^{pr}(i)$, eventually the root. Let us call $p \in \mathcal{P}$ the only path such that $s(p) = j$ and $t(p) = i$. The label $< j, C(i, j) >$ of i has a cost of $C_{D_i}^{min} \cdot \Delta_p$.

⋆ Let us consider a non-leaf node $i \in V^*$ such that all nodes in $\Gamma^+(i)$ have been labeled. In the minimal cost network, the cables going through arc $(\gamma(i), i)$ are all born in a node $j \in V^{pr}(i)$. Thanks to the maintenance constraint, we know that they are all born in the same node. Since all nodes $k \in \Gamma^+(i)$ have been labeled, for each of these nodes, there is a node $j_k \in V^{pr}(k)$ such that in the label $< j_k, C(k, j_k) >$, $C(k, j_k)$ describes the cost of the minimum cost network in the arborescence rooted in k plus the cost of the cables on the path from j_k to k. Furthermore, since the cables going through arc $(\gamma(i), i)$ are all born in j, we have either $j_k = j$ or $j_k = i$. Let us consider the label $<j, C(i, j)>$ of node i. If in the minimal network i is module-served, then we will have $u = 0$ in the computation of (26). Furthermore, let us consider $k \in \Gamma^+(i)$. If $j_k = i$, we will have $k \in S$ in the computation of (26), and $k \in \Gamma^+(i) \setminus S$ otherwise. Hence the result.   □

The termination of the algorithm derives from Proposition 6. For each node $r' \in \Gamma^+(r)$, we have $V^{pr}(r') = \{r\}$. This implies, using this proposition, that in the label $< r, C(r', r) >$, $C(r', r)$ is the cost of the minimum network cost in the arborescence rooted in r' plus the cost of the cables on (r, r'). Summing these values gives the minimum network cost.

The computation of (26) at each step is not done in polynomial time. There are many algorithms able to tackle it (dynamic programming, brute search, ...). We propose a way to tackle it in the next section which allows us to give an approximation in polynomial time, thus providing a polynomial time approximation algorithm.

4.2   Approximation Algorithm

In this Section, we propose here a Fully Polynomial Time Approximation Scheme (FPTAS) for ESFCND, in the case where:

– The height of the arborescence describing the civil engineering is upper bounded by $H \in \mathbb{N}$.
– The number of intervals on which the cost of the welds PW is a linear function with respect to m is upper bounded by $F \in \mathbb{N}$ (recall that PW is defined to be piecewise linear).

We introduce the following additional notation. PW is decomposed into its linear components. For $f \in \{1, .., F\}$, we have successive integers B_f such that $\forall m \in \{B_f, .., B_{f+1}\}, PW_m = PW^{a,f} \times m + PW^{b,f}$.

A FPTAS for the knapsack problem is available in [11]. This algorithm $\mathcal{A}$ gives, for an instance of the knapsack problem, and a number $\alpha > 1$, a solution S to the knapsack problem of cost C^{approx} where $C^{approx} \leq \alpha \times OPT$ and OPT is the optimal solution cost (here, we consider the minimization version of the knapsack problem, or "covering problem").

In Algorithm 1, the computation of (26) is the only step which is not done in polynomial time. We propose to solve it with Algorithm 2, which reformulates it as a series of knapsack problems. Then, each of the knapsack problems can be approximated thanks to the knapsack FPTAS.

The algorithms spans all possible cable sizes. For each cable size l, it computes the minimum cost splicing operation in which a cable of size l is spliced in i. (27) computes the minimal cost splicing in the case $u = 0$, and (28) computes the minimal cost splicing in the case $u = 1$. Finally, in line 15, it compares the best splicing obtained with the cost of continuing all cables.

The following lemma stems from the concavity of PW.

Lemma 2. $\forall (f, f') \in \{1, .., F\}^2$, if $f \leq f'$, then $\forall m \geq B_{f'}, PW^{a,f'} \times m + PW^{b,f'} \leq PW^{a,f} \times m + PW^{b,f}$

Proof. Let us assume $\exists (f_1, f_2) \in \{1, .., F\}^2$, with $f_1 \leq f_2$ and $\exists m \geq B_{f_2}$ such that $PW^{a,f_2} \times m + PW^{b,f_2} > PW^{a,f_1} \times m + PW^{b,f_1}$.

Since $PW^{b,f}$ is decreasing with respect to f, this means $PW^{a,f_2} > PW^{a,f_1}$, which contradicts the concavity of PW.

Hence the result. $\square$

Algorithm 2. Computation of (26).

1: **procedure** $C(i, j)$ CALCULATION()
2:     Define Cmin $:= +\infty$
3:     **for** $l \in \mathcal{L}$ **do**
4:         **for** $f \in \{1, .., F\}$ **do**
5:             **if** $(\{M_{l-1} + 1, .., M_l\}) \cap \{B_f, .., B_{f+1}\} \neq \emptyset$ **then**
6:                 $m_1 := \max(M_{l-1} + 1, B_f)$
7:                 Solve the following knapsack problems

$$C_1 = \min \sum_{k \in \Gamma^+(i)} \left(x_k \cdot C(k,i) + (1 - x_k) \cdot C(k,j) \right)$$
$$+ PW^{a,f} \times \sum_{k \in \Gamma^+(i)} x_k \cdot m_{i,k}^{act} + PW^{b,f} + \Delta_{(i,j)} \cdot C_l^{le}$$
$$+ \Delta_{(i,j)} \cdot C^{min}(D_i) + PB_l \tag{27}$$

such that $\sum_{k \in \Gamma^+(i)} x_k \cdot m_{i,k}^{act} \geq m_1$

8:             **end if**
9:             **if** $(\{M_{l-1} + 1 - D_i, .., M_l - D_i\}) \cap \{B_f, .., B_{f+1}\} \neq \emptyset$ **then**
10:                 $m_2 := \max(M_{l-1} + 1 - D_i, B_f)$

$$C_2 = \min \sum_{k \in \Gamma^+(i)} \left(x_k \cdot C(k,i) + (1 - x_k) \cdot C(k,j) \right)$$
$$+ PW^{a,f} \times \sum_{k \in \Gamma^+(i)} x_k \cdot m_{i,k}^{act} + PW^{b,f} + \Delta_{(i,j)} \cdot C_l^{le} + PB_l \tag{28}$$

such that $\sum_{k \in \Gamma^+(i)} x_k \cdot m_{i,k}^{act} \geq m_2$
                 $x \in \{0, 1\}^{|\Gamma^+(i)|}$

11:             **end if**
12:             $C^{min} := \min(C^{min}, C_1, C_2)$
13:         **end for**
14:     **end for**
15:     $C^{min} := \min(C^{min}, \sum_{k \in \Gamma^+(i)} C(k,j))$
16:     **return** C^{min}
17: **end procedure**

From this lemma, we can get that if, for some $l \in \mathcal{L}$ and $f \in \mathcal{F}$, C_1 is reached for values of x_k such that $\sum_{k \in \Gamma^+(i)} m_{i,k}^{act} x_k > B_{f+1}$ (the values returned by the knapsack problem are higher than the range of welds we consider), then a lower value of C_1 can be reached for l and $f + 1$. A similar reasoning can be done for C_2.

Let us consider $H \in \mathbb{N}$. Let us consider an instance of ESFCND where the civil engineering arborescence height is upper bounded by a constant H. We propose the following FPTAS for ESFCND.

Let us consider $\alpha > 1$. There is a polynomial time algorithm $\mathcal{A}$ which approximates the knapsack within a ratio $\alpha^{\frac{1}{H}}$. Run algorithm $\mathcal{A}'$ which is a variant of algorithm 1 where:

- Each computation of (26) is done with Algorithm 2.
- In Algorithm 2, each computation of (27) and (28) is approximated with algorithm $\mathcal{A}$.

This algorithm runs in polynomial time. Indeed, in Algorithm 1, the only step which is not done in polynomial time is replaced by a polynomial time algorithm.

Proposition 7. *Algorithm $\mathcal{A}'$ returns a cost v of the ESFCND problem such that $v \leq \alpha v^*$ where v^* is the cost of optimal solution of ESFCND.*

Proof. ⋆ Let us consider a leaf node $i \in V_D$. The labels $C(i, j)$ for $j \in V^{pr}(i)$ have the same value in Algorithm 1 and algorithm $\mathcal{A}'$.

⋆ Let us consider a non-leaf node $i \in V^*$ and $j \in V^{pr}(i)$. In the computation of (26) by Algorithm 2, C^* is approximated with a ratio of $\alpha^{\frac{1}{H}}$. Its value is the sum of welds and boxes costs and of a linear combination of the values of $C(k, i)$ and $C(k, j)$ for $k \in \Gamma^+(i)$. So it multiplies the approximation ratios of the values of $C(k, i)$ and $C(k, j)$. Hence, each time a node is labeled, the approximation ratio of its labels are $\alpha^{\frac{1}{H}}$ time the approximation ratio of its children node.

Hence the global multiplicative ratio of this algorithm is α.

The next section assesses the complexity of SFCND and ESFCND.

5 Complexity

We show in Sect. 5.1 that SFCND is NP-hard even with 1 cable size and an upper bound on the node degree of 2, and in Sect. 5.2 that ESFCND is NP-hard.

5.1 SFCND

Upper Bounded Degree. Let us consider the Number Partitioning Problem (NPP), which is shown to be NP-complete in [10].
Instance: A set of N strictly positive integers $\{n_i \in \mathbb{N} | i \in \{1, .., N\}\}$.
Question: Is there a partition of the integers $S \subseteq \{1, .., N\}\}$ such that $\sum_{i \in S} n_i = \sum_{i \notin S} n_i$?

We consider an instance of the NPP and associate it to the following SFCND instance: Let $G = (V, A)$ be an arborescence describing the civil engineering structure, $(V = \{r, 0, 1\} \cup \{v_i | i \in \{1, .., N\}\}$, $A = \{(r, 0); (0, 1); (1, v_1); (v_{i-1}, v_i) | i \in \{2, .., N\}\})$ (G is a chain graph), r is the fiber source. The demand nodes are

$\{v_i, i \in \{1, .., N\}\}$ and have respective demands $n_i, i \in \{1, .., N\}$ modules. Only one type of cable is available, with size $M_1 = \frac{1}{2} \sum_{i \in \{1,..,N\}} n_i$. Its cost per length unit is $C_1 = 1$. The lengths of all arcs of the arborescence are null, except $(r, 0)$ which is of length 1. This means the cost of a cable born in r is 1, and the cost of the other ones is 0. The cost of welds and boxes is null.

The question associated to this SFCND instance is "Is there a cabling solution cheaper than 2 ?".

$\star$ If (NPP) is feasible: $\exists S \subseteq \{1, .., N\}$ such that $\sum_{i \in S} n_i = \sum_{i \notin S} n_i$. We then build the following cabling solution:

– Two cables holding only active modules are installed on link $(r, 0)$.
– In node 0, one incoming cable is spliced into $N - |S|$ born cables. The born cables have a number of active modules $n_i, i \notin S$ and serve respectively the demand nodes $(v_i)_{i \notin S}$.
– In node 1, the cable coming from the root with only active modules is spliced into $|S|$ born cables. The born cables have n_i active modules and serve the demand nodes $(v_i)_{i \in S}$.

Since the number of active modules is conserved in each splicing, the cabling solution described above is feasible (it is illustrated in Fig. 5, as well as the instance). Its cost is equal to 2.

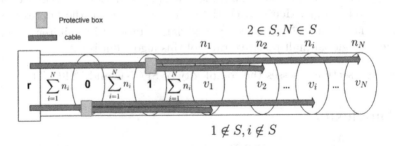

Fig. 5. Instance and solution used in the complexity proof [3].

$\star$ If (NPP) is not feasible. Then, the solution described above is not possible anymore. One cable is not large enough to cover link $(r, 0)$. Two cables cannot cover $(r, 0)$ either, since they would both have only active modules, which would mean that the (NPP) problem was feasible. Consequently, at least 3 cables need to be installed on arc $(r, 0)$, and such a solution has a cost of a least 3.

Remark 4. The solution illustrated in Fig. 5 is not valid for ESFCND, the maintenance rule is not respected in nodes 0 and 1.

Upper Bounded Arborescence Height. We show in the following that the problem is still NP-hard when restricted with:

- One cable size available.
- Civil engineering arborescence height of 3.
- Null welding cost.

We consider an instance of (NPP) that we associate to the following FCNDA instance.

Let (V, A) be an arborescence describing the civil engineering structure ($V = \{r, 0, 1\} \cup \{v_i | i \in \{1, .., N\}\}$,

$A = \{(r, 0); (0, 1); (1, v_i) | i \in \{1, .., N\}\})$; only one type of cable with a number of modules $M_1 = \frac{1}{2} \sum_{i \in \{1, .., N\}} n_i$ is available, its linear cost is $C_1 = 1$. The length of all arcs of the arborescence are zero, except $(r, 0)$ which is of length 1. This means the cost of a cable created in r is 1, and the cost of the other ones is 0. The number of active modules associated with each arc are: $m_{(r,0)}^{act} = m_{(0,1)}^{act} = \sum_{i=1}^{N} n_i; \forall i \in \{1, .., N\}, m_{(1,v_i)}^{act} = n_i$, which means that the demand points are the $v_i, i \in \{1, .., N\}$ and have respective demands n_i. This network is represented in Fig. 6. We consider a zero cost for welding and welding boxes.

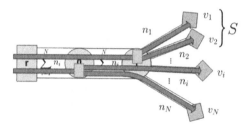

Fig. 6. Solution and instance considered in the NP-completeness proof.

The question associated to this FCNDA instance is "Is there a cabling solution cheaper than 2 ?".

Let us first assume that (NPP) is feasible: $\exists S \subseteq \{1, .., N\}$ such that $\sum_{i \in S} n_i = \sum_{i \notin S} n_i$. We then build the following cabling solution:

- Two cables holding only active modules are installed on link $(r, 0)$.
- In node 0, one incoming cable is spliced into $N - |S|$ born cables. The born cables have a number of active modules $n_i, i \notin S$ and serve the demand nodes $(v_i)_{i \notin S}$.
- On link $(0, 1)$, one cable coming from r with only active modules, and $N - |S|$ cables serving demand nodes in $\{v_i | i \notin S\}$ are installed.
- In node 1, the incoming cable with only active modules is spliced into $|S|$ born cables. The born cables have n_i active modules and serve the demand nodes $(v_i)_{i \in S}$.
- One cable is installed on each link $(1, v_i)$.

Since the number of active modules is conserved in each splicing, the cabling solution described above is feasible (it is illustrated in Fig. 6). Its cost is equal to 2, as the cables created in r have a cost of 1, and the other ones have a cost of 0.

Inversely, let us assume that (NPP) is not feasible: then, the solution described above is not possible anymore. One cable is not large enough to cover link $(r, 0)$, it cannot contain all the required active modules. Let us assume there is a solution with only two cables on $(r, 0)$. Since their combined number of modules is $\sum_{i \in \{1,..,N\}} n_i$, they both hold only active modules. If one of them directly served the demand without enduring any operation, then the (NPP) instance was trivially feasible (one of the n_i is half the total sum). So both of them endure a splicing operation, one in node 0, the other in node 1. Let us consider the cables created in 1. They serve a subset S_1 of the demand nodes, and have a respective number of active modules of $n_i, i \in S_1$. Since the number of active modules in a splicing operation is conserved, we have $\sum_{i \in S_1} n_i = \frac{1}{2} \sum_{i \in \{1,...,N\}} n_i$ and the (NPP) instance was feasible.

Consequently, at least 3 cables need to be installed on arc $(r, 0)$, and such solution has a cost of at least 3.

5.2   ESFCND

ESFCND can be shown to be NP-complete by reduction from the (NPP). With the same notations, let us consider an instance of the NPP and associate it to the following ESFCND instance. The civil engineering structure is described by the set of nodes is $V = \{r, 0\} \cup \{v_i | i \in \{1, .., N\}\}$; the set of arcs $A = \{(0, v_i) | i \in \{1, .., N\}\} \cup \{(r, 0)\}$; r is the fiber source, the nodes $\{v_i | i \in \{1, .., N\}\}$ have a demand of n_i modules. The length of all arcs except $(r, 0)$ is null. We have $N + 1$ cables available:

- N cables of sizes n_i modules and cost per length unit n_i
- A cable of size $\frac{1}{2} \sum_{i=1}^{N} n_i$ and cost per length unit $\frac{1}{2} \sum_{i=1}^{N} n_i - 1$

The cost of welds and boxes is null.

The question we ask is "is there a solution of cost at most $\sum_{i=1}^{N} n_i - 1$"?

⋆ If (NPP) is feasible. Then, we have $S \subseteq \{1, .., N\}$ such that $\sum_{i \in S} n_i = \sum_{i \notin S} n_i$. We consider the solution of ESFCND where

- For $i \in \{1, .., N\}$, on each arc $(0, v_i)$, we lay down a cable of size n_i
- In the node 0, a cable of size $\frac{1}{2} \sum_{i \in \{1,..,N\}} n_i$ is spliced. Cables of size $n_i, i \in S$ are born, and serve the demand of nodes $v_i, i \in S$.
- On the arc $(r, 0)$, a cable of size $\frac{1}{2} \sum_{i \in \{1,..,N\}} n_i$ holding only active modules is deployed (the one spliced in 0); as well as $N - |S|$ cables of sizes $n_i, i \notin S$ which serve the demand in nodes $v_i, i \notin S$.

The cost of this solution is the cost of cables on arc $(r, 0)$ which is $\sum_{i \in \{1,..,N\}} n_i - 1$. It is illustrated in Fig. 7.

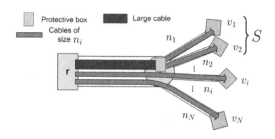

Fig. 7. Instance and solution used in the complexity proof for ESFCND [3].

$\star$ If (NPP) is not feasible. In a minimal cost solution, the size of cables serving the demand is known. For a given $i \in \{1, .., N\}$, v_i is served by a cable of size n_i. Which leaves three types of solutions to consider.

The solution without splicing has a cost $\sum_{i \in \{1,..,N\}} n_i$. Each demand node is served by a cable coming directly from the root r.

Any solution where a cable of size $\frac{1}{2} \sum_{i \in \{1,..,N\}} n_i$ is spliced in 0 has a cost at least equal to $\sum_{i \in \{1,..,N\}} n_i$. Indeed, let us note $E \subseteq \{1, .., N\}$ the set such that cables of sizes $n_i, i \in E$ are born in 0. Since the NPP instance is not feasible, we have $\sum_{i \in E} n_i < \frac{1}{2} \sum_{i \in \{1,..,N\}} n_i$, so the cost of cables which are continued in 0 is $\sum_{i \notin E} n_i > \frac{1}{2} \sum_{i \in \{1,..,N\}} n_i$, and the total cost of the network is $\sum_{i \notin E} n_i + \frac{1}{2} \sum_{i \in \{1,..,N\}} n_i - 1 \geq \sum_{i \in \{1,..,N\}} n_i$.

Any solution where a smaller cable is spliced in 0 has a cost at least equal to $\sum_{i \in \{1,..,N\}} n_i$. Indeed, in any splicing of a cable of size n_i for a given $i \in \{1, .., N\}$, the spliced cable is at least as expensive than the born cables.

5.3   Synthesis

To the results proven here, we can add those deducible from Sect. 4. The restriction of ESFCND where there is an upper bound on the node degree can be solved in polynomial time, since in that case the computation of (26) can be done in polynomial time. This implies that it is also polynomial when more parameters are fixed. Furthermore, we showed in Sect. 4.2 that the problem admits a FPTAS under some conditions. As for SFCND, its NP-hardness in a restricted setting implies its NP-hardness in the more general cases. These results are summed up in Table 1.

Table 1. Complexity of the two problems in different contexts.

Problem	Complexity			
Fixed elements	None	Maximum degree	Maximum degree L	F, H
SFCND	NP-hard	NP-hard	NP-hard	NP-hard
ESFCND	NP-hard	P	P	NP-hard, FPTAS

Table 1 shows a theoretical difference in the complexities of the two problems ESFCND and SFCND. We assess the numerical aspect of this difference in the next section.

6   Results

We assessed the solution methods on real-life instances taken from the city of Arles (France).

The cables available have a size of 1, 2, 4, 6, 8, 12, 18 or 24 modules. The resolution algorithm for the MIPs was the Cplex 12.6 default branch-and-bound algorithm.

6.1   Models Comparison

The results of the numerical experiments regarding the SFCND and ESFCND problem are displayed respectively in Tables 4 and 5, "base model" always refers to the MIP without valid inequalities, and "enhanced model" to the MIP with valid inequalities. The columns of both tables are labeled as follows: "time" stands for the computation time; "CR" stands for the continuous relaxation as a ratio of the optimal solution; "Br" stands for the number of explored branches of the Branch and Bound algorithm.

Regarding SFCND, the valid inequalities have had a positive effect on the average computation time, which went down from 546 to 62 s. However, on most instances (8 out of 9), the MIP is solved faster without the valid inequalities. This suggest that they are more useful for instances that are hard to solve. Regarding the algorithm, the continuous relaxation goes from an average of 90.5% to 92.6%. The high relaxation of the base model can explain the mitigated impact of the inequalities on the performances (Table 2).

Regarding ESFCND, all instances were easier to solve (computation times are displayed in milliseconds). The valid inequalities have had a beneficial effect on the computation time, all instances are solved faster with the enhanced formulation. The average computation time goes from 1730 to 329 ms. On an algorithmic level, the initial relaxation goes from an average of 13.2% of the optimal solution cost to 87.3% of the optimal solution cost. This has a significant impact on the number of nodes of the branch-and-bound algorithm, which goes from an average of 1100 branches to an average of 4 branches; 7 instances out of 9 were

Table 2. Key features of the real-life instances.

Instance	Features			
	Max degree	Arcs	Demand nodes	Total demand
Ar 1	4	113	45	61
Ar 2	6	103	38	55
Ar 3	5	103	35	66
Ar 4	6	123	43	80
Ar 5	7	129	44	68
Ar 6	6	137	43	67
Ar 7	4	139	35	68
Ar 8	5	163	41	63
Ar 9	4	219	68	78

Table 3. Key features of the fictive instances.

Instance	Features			
	Max degree	Arcs	Demand nodes	Total demand
Fi 10	11	20	15	71
Fi 11	12	22	16	84
Fi 12	13	24	18	97
Fi 13	14	26	19	112
Fi 14	15	28	21	112
Fi 15	16	30	22	127
Fi 16	17	32	24	144

Table 4. Results for SFCND.

Instance	Base formulation		Enhanced formulation	
	Time (s)	CR (%)	Time (s)	CR (%)
Ar 1	8	90.3	16	91.0
Ar 2	9	83.7	24	92.4
Ar 3	17	92.2	22	93.3
Ar 4	19	89.2	46	90.0
Ar 5	1	94.9	2	95.2
Ar 6	2	92.5	3	94.7
Ar 7	13	92.4	29	93.7
Ar 8	8	89.6	12	91.7
Ar 9	4837	89.4	408	91.6

solved without branching. The exact dynamic programming approach was more efficient than the enhanced integer programming formulation, it solved 7 out of 9 instances faster. The approximated algorithm was run with an approximation ratio of 2. It was longer than Algorithm 1 on 8 instances out of 9. Despite their similar structure, this can be explained by additional loops in the approximation algorithm, which can increase its computation time.

6.2 Sensitivity Analysis

Section 5 points to the maximal node degree as a key element of the problems complexity. Since the highest node degree of all real-life instances is between

Table 5. Results for ESFCND.

Instance	Base formulation			Enhanced formulation			Dynamic programming	FPTAS
	Time (ms)	CR (%)	Br	Time (ms)	CR (%)	Br	Time (ms)	Time (ms)
Ar 1	1457	14.0	1191	305	89.2	0	324	454
Ar 2	1174	17.8	462	239	86.6	0	239	141
Ar 3	1317	13.6	153	318	81.7	0	66	203
Ar 4	742	15.7	72	268	86.8	0	87	168
Ar 5	746	18.2	0	477	89.2	0	88	120
Ar 6	1477	15.5	66	238	91.8	0	110	235
Ar 7	1667	9.7	1045	190	80.1	0	121	251
Ar 8	1786	9.4	414	344	89.8	21	103	121
Ar 9	5204	5.3	6302	507	90.8	9	306	446

Table 6. Computation time on fictive instances (ms).

Instance	Enhanced model SFCND	Enhanced model ESFCND	Dynamic programming	FPTAS
Fi 10	205	166	322	16
Fi 11	327	77	652	17
Fi 12	993	332	1409	19
Fi 13	1130	120	3800	15
Fi 14	1369	347	12 403	28
Fi 15	1450	98	39 654	38
Fi 16	2691	280	164 243	52

4 and 7, we used fictive instances to assess the performances of each resolution technique when some of the nodes have a high degree. Their features are displayed in Table 3.

As expected, the dynamic programming algorithm was very sensitive to the node degree, the computation time growing exponentially (it was multiplied by over 500 between the smaller and larger instance). On the other hand, the approximation algorithm was much less sensitive to the node degree, with an average computation time of 26 ms. There was a smaller growth on the instances considered (it was multiplied by less than 4 between the smaller and larger instance). The enhanced MIP formulation for ESFCND was able to solve all instances in less than one second, with an average of 200 ms. This is the opposite of the results obtained on real-life instances, where the dynamic programming was more efficient. As for SFCND, the MIP formulation proved to be efficient, with an average computation time of 900 ms. Although the instances with a higher degree are harder to solve, it stays tractable in practice. One should favor a MIP based approach, regardless of the problem, when dealing with high degree nodes (Table 6).

6.3   Operational Considerations

We compared the optimal solutions of both problems, as well as the approximated solutions found. The approximation ratio selected was still of 2. Results are displayed in Table 7, the column labeled "arcs with rule broken" denotes the number of arcs where the maintenance rule (illustrated in Fig. 4) is broken.

Table 7. Optimal solution costs and characteristics.

Instance	Approximated solution ESFCND	Optimal solution ESFCND	Optimal solution SFCND	Arcs with rule broken
Ar 1	6156.6	6156.6	6087.3	6
Ar 2	10 382.1	10 357.3	9870.0	8
Ar 3	6568.6	6546.2	6125.8	14
Ar 4	6788.1	6720.8	6461.9	14
Ar 5	5081.8	5081.8	5081.8	0
Ar 6	6546.5	6546.5	6544.2	1
Ar 7	9734.6	9348.0	8638.6	18
Ar 8	12 328.3	12 328.3	12 248.4	4
Ar 9	26 309.7	25 619.1	24 422.8	15

The ESFCND solutions provided by the approximation algorithm were in average 1% more expensive than the optimal solutions, with the two being equal for 4 instances out of 9. This can be seen as a good performance, and is much better than the worst case guarantee.

The optimal solution of ESFCND is in average 3.7% more expensive than the optimal solution of SFCND. This can be seen as an acceptable loss in capital expenditure if it is compensated by an easier maintenance, depending on the importance accorded to it.

The maintenance rule is broken in almost every real-life instance we tried (8 out of 9). In average, it is not respected in 6.2% of the arcs, which is significant. This suggests that the optimal solutions of SFCND will be much harder to repair in case of failure on one of the arcs. These elements can be taken into account to establish a strategy in case of node failure.

7   Conclusion

This chapter tackles two fiber cables network design problems, one unconstrained by maintenance consideration (SFCND) and the other one constrained (ESFCND). Regarding the unconstrained problem, one integer programming based solving algorithm was proposed. Associated valid inequalities make it more tractable in practice. We proposed two exact solution methods for the constrained problem. These methods are complementary, as they prove efficient in different contexts: the dynamic programming approach is generally faster in graphs where nodes have a small degree, whereas the mixed integer programming, embedding efficient valid inequalities, is generally faster otherwise. An FPTAS was also provided, which was faster in both cases, while providing good quality solutions.

On a theoretical level, the unconstrained problem seems much more complex to solve than the constrained problem. Fixing some parameters makes the constrained problem polynomial, or approximable, while the unconstrained problem stays NP-hard. Our numerical experiments confirmed this tendency on real-life instances.

As for the operational side, the maintenance rule can be considered as a reasonable compromise between capital expenditure for the network deployment and maintenance costs. Its implementation only increases the optimal solution cost by 3.7% on our test instances.

References

1. Angilella, V., Chardy, M., Ben-Ameur, W.: Cables network design optimisation for the fiber to the home. In: Design of Reliable Communication Networks (2016)
2. Angilella, V., Chardy, M., Ben-Ameur, W.: Design of fiber cable tree networks for the fiber to the home. In: International Networks Optimisation Conference, Lisboa, Portugal (2017)
3. Angilella, V., Chardy, M., Ben-Ameur, W.: Fiber cable network design with operations administration & maintenance constraints. In: International Conference on Operations Research and Enterprise Systems, Funchal, Madeira, Portugal (2018)
4. Bley, A., Ljubic, I., Maurer, O.: Lagrangian decompositions for the two-level FTTx network design problem. Eur. J. Comput. Optim. 1(3), 221–252 (2013)
5. Chardy, M., Costa, M.C., Faye, A., Trampont, M.: Optimising splitter and fiber location in a multilevel optical FTTH network. Eur. J. Oper. Res. 222(3), 430–440 (2013)
6. Contreras, I., Fernandez, E.: General network design: a unified view of combined location and network design problems. Eur. J. Oper. Res. 219(3), 680–697 (2012)
7. FTTH Council Europe: FTTH Handbook, 7th edn. Wettelijk Depot (2016)
8. Gollowitzer, S., Gouveia, L., Ljubic, I.: Enhanced formulations and branch and cut for the two level network design problem with transition facilities. Eur. J. Oper. Res. 2, 211–222 (2013)
9. Grötschel, M., Raack, C., Werner, A.: Towards optimising the deployment of optical access networks. Eur. J. Comput. Optim. 2(1–2), 17–53 (2013)

10. Karp, R.M.: Reducibility among combinatorial problems. In: Miller, R.E., Thatcher, J.W., Bohlinger, J.D. (eds.) Complexity of Computer Computations. The IBM Research Symposia Series. Springer, Boston (1972). https://doi.org/10.1007/978-1-4684-2001-2_9

11. Magazine, M., Oguz, O.: A fully polynomial approximation algorithm for the 0–1 knapsack problem. Eur. J. Oper. Res. **8**(3), 270–273 (1981)

12. Mateus, G.R., Luna, H.P., Sirihal, A.B.: Heuristics for distribution network design in telecommunication. J. Heuristics **6**, 131–148 (2000)

Impact of Iterated Local Search Heuristic Hybridization on Vehicle Routing Problems: Application to the Capacitated Profitable Tour Problem

Hayet Chentli[1(✉)], Rachid Ouafi[1], and Wahiba Ramdane Cherif-Khettaf[2]

[1] Department of Operations Research, USTHB, P.O. Box 32 El Alia,
16111 Bab Ezzouar, Algiers, Algeria
chentli.hayet@gmail.com, hchentli@usthb.dz
[2] LORIA, UMR 7503, Lorraine University, Mines Nancy, Nancy, France

Abstract. The present paper highlights the impact of heuristic hybridization on *Vehicle Routing Problems* (VRPs). More specifically, we focus on the hybridization of the *Iterated Local Search heuristic* (ILS). We propose different hybridization levels for ILS with two other heuristics, namely a *Variable Neighborhood Descent with Random neighborhood ordering* (RVND) and a *Large Neighborhood Search heuristic* (LNS). To evaluate the proposed approaches, we test them on a variant of VRPs called the *Capacitated Profitable Tour Problem* (CPTP). In a CPTP, the visit of all customers is no longer required and the visit of each customer generates a specific profit. The available fleet of vehicle is limited and capacitated. The aim of the CPTP is to choose which set of customers to visit and in which order to maximize the difference between collected profits and routing costs. Our experiments show that the more ILS is hybridized the better are the results. To bring out the effectiveness of the proposed hybrid approach combining ILS, RVND and LNS, a comparison is made between that proposed approach and three local search heuristics from the literature of the CPTP. The obtained results are competitive.

Keywords: Heuristics · Hybridization · Vehicle Routing Problem · Iterative local search

1 Introduction

In recent years, considerable attention has been paid to logistic problems in general and to *Vehicle Routing Problems* (VRPs) in particular. Different methodologies have been adopted to "solve" that kind of problems. Among the proposed methodologies, heuristic algorithms are particularly much studied. Researchers in the fields of combinatorial optimization are trying their best to improve the solution quality and the computing time of previously proposed heuristics, especially when it comes to solve difficult problems as VRPs.

© Springer Nature Switzerland AG 2019
G. H. Parlier et al. (Eds.): ICORES 2018, CCIS 966, pp. 80–101, 2019.
https://doi.org/10.1007/978-3-030-16035-7_5

The present work aims at analyzing the impact of heuristic hybridization on VRPs. More specifically, hybridizations of the *Iterated Local Search heuristic* (ILS) with other single-solution based heuristics are considered.

We recall that ILS principle is to improve a given initial solution by alternating *local search* (LS) and *perturbation procedures*. The role of a LS is to improve a given solution by performing a set of small modifications (or moves) to the studied solution. One can say that the LS visits the neighborhood of the studied solution and selects the best neighboring solution according to some criterion. After some iterations, the LS is no longer able produce better quality solutions using the same set of moves. We say that the heuristic is trapped in a local optimum.

To help LS to escape local optima, ILS provides a perturbation procedure. The latter procedure performs some changes to the current local optimum, producing thereby a new starting solution for the LS. The quality of that new starting solution is generally not as good as the quality of the local optimum. That decrease in the solution quality induced by the perturbation procedure allows the LS to visit a larger search space area.

In the present paper, we attempt to improve a simple ILS heuristic by modifying its local search and perturbation procedures. Several ILS hybridizations are implemented based on several simple LS heuristics, a *Variable Neighborhood Descent with Random neighborhood ordering* (RVND) and a *Large Neighborhood Search heuristic* (LNS).

To assess the performance of the studied heuristics, the latter are tested on a VRP variant called *Capacitated Profitable Tour Problem* (CPTP).

The CPTP has been introduced by Archetti et al. [1] in order to deal with empty returns, that trucks are facing after performing delivery operations (see [1] for more details). The main difference between the CPTP and the classical VRP is the relaxation of the constraints imposing a visit for each customer. In addition, in a CPTP a profit is assigned to each customer and the objective is to maximize the difference between collected profits and routing costs. The number of available vehicles in a CPTP is supposed to be finite. These vehicles are homogeneous with a fixed capacity bound. Each customer in a CPTP has a given pickup demand that must be entirely fulfiled if the customer is visited. Furthermore, if a customer is included in the solution, its demand has to be satisfied by performing a single visit.

The rest of the paper is organized as follows. Section 2 presents some previous works from the literature dealing with the application of ILS to combinatorial optimization problems and VRPs. Hybridizations of ILS with other heuristics are highlighted. A CPTP literature review is also given in that Section. Section 3 describes the proposed approaches. Section 4 discusses the computational results. Finally, a conclusion is given in Sect. 5.

2   Literature Review

The present section is divided into two subsections. In the first subsection, we provide the literature review of the Iterated Local Search (ILS) heuristic. While the second subsection is devoted to the CPTP heuristic approaches proposed in the literature.

2.1   Iterated Local Search Heuristic

According to Lourenço et al. [2], ILS is an efficient heuristic that has several desirable features of a metaheuristic. The main features are the simplicity, the high effectiveness, the robustness and the ease and the malleability of implementation (several implementation choices are left to the developer). The authors also state that ILS effectiveness depends on the choice of the used modules: local search, perturbation procedure and acceptance criterion.

In the literature, several researchers attempted the resolution of combinatorial optimization problems using hybridization of ILS with other heuristics. For instance, Martins et al. [3] developed a *Variable Neighborhood Descent* (VND) combined with an ILS for the Routing and Wavelength Assignment problem. In that paper, VND plays the role of the local search procedure and uses three neighborhood structures. When VND is blocked, ILS perturbs the so far obtained solution and the process iterates until a stopping criterion is met.

Martins et al. [4] implemented a hybrid ILS and RVND heuristic for the Cell Formation Problem. The proposed RVND uses three neighborhood structures. In addition, three perturbation procedures are used in ILS.

Many researchers successfully applied hybrid ILS heuristics to VRP variants. For example, Chen et al. [5] developed a hybridization of ILS with VND for the Capacitated Vehicle Routing Problem. VND uses two inter- and two intra route(s) operators consisting of intra-route relocation, 2-opt, inter-routes swap and 2-opt*. The perturbation phase is performed using the cross-exchange operator.

Subramanian et al. [6] proposed a parallel algorithm combining an ILS with a RVND for solving the Vehicle Routing Problem with Simultaneous Pick-up and Delivery services. Five intra- and seven inter-route(s) neighborhood structures are given together with three perturbation mechanisms.

Subramanian et al. [7] implemented a hybrid algorithm combining an exact method with ILS and RVND for a class of VRPs with heterogeneous fleet. The ILS and RVND heuristics are based on those presented in [6].

Assis et al. [8] presented a hybrid ILS using RVND in the local search phase. The proposed RVND uses six inter- and six intra-route(s) neighborhood structures. That hybrid approach is tested on the multiobjective vehicle routing problem with fixed delivery and optional collections.

Another hybridization of ILS is implemented by Subramanian and Battarra [9] to solve the Travelling Salesman Problem with Pickups and Deliveries. The authors hybridized ILS with RVND. In RVND, four neighborhood structures are given.

Hernández-Pérez et al. [10] studied a hybridization of ILS with VND. The approach is applied to the multi-commodity Pickup-and-Delivery Traveling Salesman Problem. The approach is tested with up to six neighborhood structures and a combination of three shaking procedures.

Todosijević et al. [11] developed another hybrid approach using both ILS and VND for the Swap-Body Vehicle Routing Problem. The used neighborhood structures are 1-Opt, 2-Opt, Or-Opt, relocate and exchange. The shaking procedure is based on customer relocation.

The ILS and the VND heuristics also provide good quality solutions for other variants of VRPs see [12–14]. In addition, the two heuristics perform well on some *Vehicle Routing Problems with Profits* (see [15]). Furthermore, several versions of VND are used to solve different variants of transportation problems (see [16–18]).

As one can see from the literature review, several papers use combinations of ILS and RVND for solving VRP variants. However, to the best of our knowledge, only one work has been addressed using a hybridization of ILS with LNS [19]. ILS and LNS heuristics are nevertheless quite effective in solving VRP variants as well as other transportation problems. For ILS, we refer the reader to the papers [20–23], and for LNS, we refer the reader to the papers [24–28].

2.2   Capacitated Profitable Tour Problem

Despite its importance, the CPTP has not received a lot of attention from researcher. Archetti et al. [1] introduced the CPTP and proposed three methodologies to solve that problem. The proposed methodologies are the Tabu Feasible (TF), the Tabu Admissible (TA) and the Variable Neighborhood Search (VNS) heuristics. Both TF and TA algorithms use two inter-route operators. The first operator is called 1-move. 1-move either relocates a given customer in a different route or deletes that customer completely from the solution. The second movement is called swap-move. Swap-move either exchanges the positions of two given customers from two different routes or deletes a customer and replaces it by an unrouted one.

In order to deal with infeasible solutions obtained by the TA algorithm, Archetti et al. proposed a repair heuristic based on series of 1-move. In addition, the authors evaluated the solutions according to several criteria including the difference between total profit and total distance, the number of routes, the route duration and the maximum constraint violation.

The VNS algorithm uses the TF algorithm with a small iteration number, which allows the visit of a larger area within the search space.

Some researchers proposed exact methods for the CPTP. As the present work deals with heuristic approaches, we do not describe those exact methods.

3   The Proposed Methodology

In the present section, we describe the implemented construction heuristic and the studied approaches: ILS, LNS and RVND. We also describe the tested hybridizations.

Note that a preliminary work dealing with ILS hybridization has been presented in [29]. The present work extends the one proposed in [29] by providing: (i) other heuristic approaches combining ILS with other neighborhood operators, (ii) a hybridization of ILS with both LNS and a neighborhood operator, (iii) a detailed comparison between the use of the *basic greedy* heuristic and a *random insertion* heuristic within the perturbation procedure, (iv) detailed results of the hybrid heuristic combining ILS, RVND and LNS compared with Archetti et al. [1] results.

3.1 Construction Heuristic

The implemented construction heuristic is a sequential heuristic based on the I1 heuristic of Solomon [30]. I1 was first developed for the Vehicle Routing Problem with Time Window. The pseudo-code of the construction heuristic is displayed in Algorithm 1.

Algorithm 1. Construction heuristic for the CPTP.

1: **Inputs:**
    A CPTP instance
    A list L_{unr} containing all the unrouted customers
    A number $nbRoutes = 0$ of the current solution routes
2: **Outputs:**
    A feasible solution
3: **while** $nbRoutes <$ vehicle number **do**
4:     Generate a new route;
5:     $nbRoutes + +$;
6:     Add a seed customer to the new route;
7:     **while** $\exists u \in L_{unr}$ whose insertion leads to a feasible solution **do**
8:         Evaluate the insertion of each unrouted customer $u \in L_{unr}$ into the studied route r_{stu};
9:         Choose the best insertion position for each u using the criterion $cr_1(i, u, j)$;
10:        Choose the customer u^* that has the best value of $cr_1(i, u, j)$;
11:        Insert the customer u^* in its best insertion position within r_{stu};
12:        Delete u^* from L_{unr};
13:     **end while**
14: **end while**

An empty route is considered in the first iteration of the construction heuristic. That empty route is first filled with a seed customer which is randomly chosen. After that, the heuristic evaluates the insertion of the remaining unrouted customers into the route. The best insertion position of each unrouted customer u between customers i and j is selected. This is done according to a given criterion denoted $cr_1(i, u, j)$. Among all the best insertion positions, the construction heuristic chooses the one that optimizes the given criterion. That process iterates until no customer can be inserted into the current route. If some customers are still unrouted, a new route is generated and the process is repeated. The heuristic stops either if there are no more unrouted customers or if the number of generated routes exceeds the vehicle number.

To describe $cr_1(i, u, j)$, let us consider $(i_0, i_1, \ldots, i_h)$ as the current route where i_ρ stands for the ρth position in the route if $\rho \notin \{0, h\}$, i_ρ stands for the depot otherwise $(\rho \in \{0, h\})$. The best insertion position of customer u within the current route is selected according to Expressions (1)–(4) (Source [29]). In these Equations c_{ij} refers to the distance between customers i and j, pr_u refers to customer u profit, $\alpha_1, \alpha_2 \geq 0$ with $\alpha_1 + \alpha_2 = 1$ are two parameters set by the user.

$$cr_1(i(u), u, j(u)) = \max\left\{cr_1(i_{\rho-1}, u, i_\rho), \rho = 1, \ldots, h\right\}; \text{ (Source [29])} \qquad (1)$$
$$cr_1(i, u, j) = \alpha_1 \cdot cr_{11}(i, u, j) - \alpha_2 \cdot cr_{12}(i, u, j); \text{ (Source [29])} \qquad (2)$$
$$cr_{11}(i, u, j) = pr_u; \text{ (Source [29])} \qquad (3)$$
$$cr_{12}(i, u, j) = c_{iu} + c_{uj} - c_{ij}. \text{ (Source [29])} \qquad (4)$$

The Eqs. (1)–(4) have been already presented in [29] on page 117, Sect. 2.1.

Note that different values of parameters α_1 et α_2 can lead to different solutions for the CPTP.

3.2 Iterated Local Search

ILS principle is described in Sect. 1. The pseudo-code of the ILS heuristic used in the present paper is given in Algorithm 2. ILS starts from an initial solution given by the construction heuristic presented in Sect. 3.1. Then a local search procedure is executed for a given number of iterations. If the obtained solution is better than the current best one, the best solution is replaced by the obtained one. The solution is, after that, perturbed using a perturbation procedure. The process is repeated until a stopping criterion is met. In the present work, the stopping criterion stands for the completion of a given number of iterations without improvement.

Algorithm 2. Iterated Local Search.

1: **Inputs:**
    A CPTP instance
2: **Outputs:**
    The best solution found
3: Generate an initial solution;
4: **while** stopping criterion is not met **do**
5:     Execute a local search procedure;
6:     Update the best solution;
7:     Perturb the obtained solution;
8: **end while**

Local Search. As mentioned by Talbi [31], any single-solution based meta-heuristic can be used in the local search phase of an ILS.

In the present work, seven basic ILS versions are first tested. In each ILS basic version, a different neighborhood structure (or neighborhood operator) is used. The implemented neighborhood operators consist of four intra- and three inter-route(s) operators. The latter are described in what follows. Examples of neighborhood movements for each operator are given in Fig. 1. Figure 1 was first given in [29] on page 118.

2-Opt introduces two new arcs and deletes two other arcs in a given route by connecting two customers k and l and reversing the path between those customers. In Fig. 1, the arcs $(1, 4)$ and $(2, 5)$ are deleted, the arcs $(1, 2)$ and $(4, 5)$ are added, customers 1 and 2 are connected and the path $(4 - 3 - 2)$ is reversed. For maintaining the route connectivity, customers 4 and 5 are connected.

2-Opt* divides two given routes into four segments: initial and final segments. Then, the operator connects each first segment from a route with a second segment from the other route. In Fig. 1, the first route $(0-1-2-3-0)$ is disconnected into a first segment $(0 - 1)$ and a second segment $(2 - 3 - 0)$. The second route $(0 - 4 - 5 - 6 - 0)$ is disconnected into a first segment $(0 - 4 - 5)$ and a second segment $(6 - 0)$. After that, $(0 - 1)$ is connected to $(6 - 0)$ and $(0 - 4 - 5)$ is connected to $(2 - 3 - 0)$.

Intra-route 1-0 Exchange relocates a customer l into a position k within a same route. In Fig. 1, customer 2, which is in the 5th position of the route, is relocated in the second position within the same route.

Inter-routes 1-0 Exchange relocates a customer l into a position k in a different route. In Fig. 1, customer 3, which is in the third position of the first route, is relocated in the 4th position of the second route.

Intra-route 1-1 Exchange exchanges the positions of two customers within a same route. In Fig. 1, customer 2 is relocated at the position of customer 5, and customer 5 is relocated at the position of customer 2. No path is reversed.

Inter-routes 1-1 Exchange exchanges the positions of two customers from two different routes. In Fig. 1, customer 3 is relocated at the position of customer 6, and customer 6 is relocated at the position of customer 3.

Or-Opt relocates two consecutive customers (or an arc) in a different position within a same route. In Fig. 1, the arc $(1, 2)$ is relocated between the depot and customer 3.

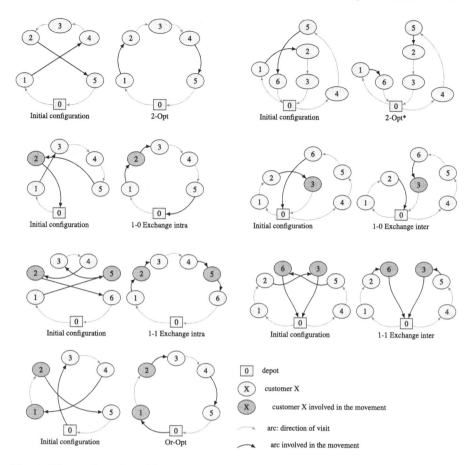

Fig. 1. Illustration of neighborhood movements in the RVND heuristic (Source [29]).

Each combination of ILS with a neighborhood operator is denoted by ILS_NeiOp where NeiOp stands for the used neighborhood operator. Thus, we have: ILS_2-Opt, ILS_2-Opt*, ILS_1-0 Exchange-intra, ILS_1-0 Exchange-inter, ILS_1-1 Exchange-intra, ILS_1-1 Exchange-inter and ILS_Or-Opt.

Perturbation Mechanism. The perturbation procedure destroys the solution obtained by the local search procedure to escape local optima. To do so, the *random removal* operator described by Pisinger and Ropke [32] is used. Before re-applying the local search procedure, some customers may be added to the obtained solution using the *basic greedy* heuristic described by the same authors [32].

3.3 Hybrid Iterated Local Search Heuristic

In the present work, several hybridization of ILS are proposed. They are described in what follows.

Hybrid ILS_LNS Heuristic. In [29], several versions of the LNS are tested. The latter versions use a unique removal and a unique insertion operator. All the removal/insertion operators proposed by Pisinger and Ropke [32] are tested using the CPTP constraints and objective function. Each implemented LNS version consists of a different couple of removal/insertion operators.

Each LNS version starts from a solution generated by the construction heuristic. At each iteration, LNS deletes a given number of customers using its removal operator. Then, LNS inserts a set of customers using its insertion operator. LNS stops when a given number of iterations without improvement is reached.

After an experimental study, we choose the *related removal* and the *regret heuristic* with a regret number equal to 4 as the couple of removal/insertion operators used in LNS.

For more information regarding the selection of the removal/insertion couple, we refer the reader to [29].

The hybrid ILS_LNS heuristic is a multi-start heuristic that executes, at each iteration, an ILS heuristic using LNS as a local search procedure. The initial solutions are obtained by iteratively modifying parameter values of the construction heuristic described in Sect. 3.1. All the possible combinations of α_1 and α_2 parameter values are considered. ILS_LNS uses the same perturbation procedure as the one described in Sect. 3.2.

As defined in Sect. 3.2, ILS stopping criterion consists in the completion of a given number of iterations without improvement.

ILS_LNS stops when all combinations of α_1 and α_2 parameter values are tested.

Hybrid ILS_RVND Heuristic. The RVND heuristic uses all the seven operators described in Sect. 3.2. At each iteration of RVND, the neighborhood operators are chosen in a random way. Actually, all the operators are put in a list of available operators, and each time an operator is used, the heuristic deletes that operator from the list. RVND stops when the list of available operators is empty. Hence, each operator is executed only once.

The hybrid ILS_RVND heuristic differs from the ILS_LNS heuristic (presented above) in the use of RVND instead of LNS in the local search phase.

Hybrid LNS_RVND Heuristic. LNS_RVND is not a multi-start heuristic. This hybrid heuristic begins with an initial solution obtained by the construction heuristic given in Sect. 3.1 using random values for parameters α_1 and α_2. Then, the LNS heuristic is run until reaching a given number of iterations. The obtained solution is possibly improved using the RVND heuristic with some probability. After that, the LNS_RVND heuristic goes back again to LNS and that process iterates until reaching a given number of iterations without improvement.

Hybrid ILS_NeiOp_LNS Heuristic. ILS_NeiOp_LNS is a multi-start heuristic that hybridizes ILS_NeiOp with LNS. NeiOp is defined in a similar manner as described in Sect. 3.2.

The neighborhood operator used in this heuristic is the one that gives the best results with respect to other neighborhood operators when combined to ILS.

In ILS_NeiOp_LNS, the construction heuristic is first run to generate different initial solutions at each iteration. Then, for each initial solution, an ILS heuristic is run until reaching a given number of iterations without improvement. In ILS, a LNS heuristic is executed followed by the selected neighborhood operator. The combination of LNS with the selected neighborhood operator plays the role of the local search procedure in ILS. This combination of LNS with the neighborhood operator is repeated for a given number of iterations. When the local search stops, the perturbation procedure is run. The latter procedure is the same as the one presented in Sect. 3.2. The pseudo-code of ILS_NeiOp_LNS is given in Algorithm 3.

Algorithm 3. ILS_NeiOp_LNS.

1: **Inputs:**
   A CPTP instance
2: **Outputs:**
   The best encountered solution
3: **while** stopping criterion is not met **do**
4:     Generate an initial solution;
5:     **while** ILS stopping criterion is not met **do**
6:         **while** stopping criterion is not met **do**
7:             Run LNS;
8:             Run NeiOp;
9:         **end while**
10:         Update the best encountered solution;
11:         Perturb the current solution;
12:     **end while**
13: **end while**

Hybrid ILS_RVND_LNS Heuristic. This hybrid heuristic is quite similar to the one presented in Sect. 3.3. The only difference is that ILS_RVND_LNS uses RVND instead of the neighborhood operator after LNS.

The pseudo-code of ILS_RVND_LNS is given in Algorithm 4.

Algorithm 4. ILS_RVND_LNS.

1: **Inputs:**
    A CPTP instance
2: **Outputs:**
    The best encountered solution
3: **while** stopping criterion is not met **do**
4:    Generate an initial solution;
5:    **while** ILS stopping criterion is not met **do**
6:        **while** stopping criterion is not met **do**
7:            Run LNS;
8:            Run RVND;
9:        **end while**
10:       Update the best encountered solution;
11:       Perturb the current solution;
12:    **end while**
13: **end while**

---

4   Computational Results

In the present Section, we begin by describing the CPTP instances proposed in the literature and studied in the present work. After that, we analyze the results of each proposed approach and we evaluate the impact of the hybridization. Finally, the approach that provides the best results is compared to some CPTP heuristics from the literature.

In order to quickly determine the best approach among the proposed ones, each approach is executed using a relatively small number of iterations. However, more iterations are used for the comparison between the best approach with the literature ones.

Note that some of the experimental results/heuristic tuning details are not given in the present work as they have already been published in the conference paper [29]. Those experimental results/heuristic tuning details concern the couple of operators chosen for LNS, the comparison between LNS and ILS_LNS and the tuning of both ILS_RVND and LNS_RVND.

The new experimentations are implemented in C and performed on a personal laptop with an Intel(R) Core (TM) i5-4210U CPU @ 1.70 GHz with 6.00 Gb RAM and 64-bit operating system.

Due to the random aspect of the approaches, they all are executed 3 times for each instance. We report the best encountered solutions in terms of percentage deviation from the best solutions presented by Archetti et al. [1]. A percentage deviation (gap) of a heuristic a from a heuristic b is computed according to the following Expression

$$gap = 100 \cdot \frac{z_b - z_a}{z_b}$$

where z_a and z_b are the objective function values obtained by heuristics a and b respectively.

4.1 CPTP Instances

The CPTP instances studied in the present work were proposed by Archetti et al. [1]. The authors modified the Capacitated Vehicle Routing Problem instances (CVRP) described by Christofides et al. [33] with 50 to 199 customers. Archetti et al. [1] generated three set of instances from the CVRP instances by varying the capacity bounds and the number of vehicles.

The first set of CPTP instances consists of the original 10 CVRP instances in which each customer i has a profit pr_i computed following the Expression $pr_i = (0.5 + h) \cdot d_i$, where d_i is the demand of i and h is randomly chosen from $[0, 1]$.

The second set of CPTP instances consists of 90 different instances obtained by modifying the first set of instances. Actually, Archetti et al. [1] consider the cases $Q = 50$, $Q = 75$ and $Q = 100$, where Q stands for to the capacity bound. For each case, three instances are generated using different vehicle numbers. The latter numbers are chosen from the set $\{2, 3, 4\}$. In the second set of CPTP instances, the profits are computed in the same manner as described for the first set.

The third set of CPTP instances consists of 30 different instances obtained by modifying the first set of instances. Archetti et al. [1] maintain the same capacity bounds as those presented for the CVRP. However, they consider three cases for the vehicle numbers. The latter are chosen from the set $\{2, 3, 4\}$.

A total of 130 CPTP instances are thus proposed by Archetti et al. [1]. As instance types $p03$ and $p08$ of Archetti et al. [1] are exactly the same, we do not consider instances of type $p03$. Hence we obtain a total of 117 CPTP instances.

4.2 Study of Basic ILS Heuristics

As said in Sect. 3.2, seven basic version of ILS are tested. Each version differs from the others in the used neighborhood operator. The tested versions are ILS_2-Opt, ILS_2-Opt*, ILS_1-0 Exchange-intra, ILS_1-0 Exchange-inter, ILS_1-1 Exchange-intra, ILS_1-1 Exchange-inter and ILS_Or-Opt.

As these basic ILS heuristics are quite fast, we decide to fix their number of iterations without improvement in ILS to 500 instead of 50.

Table 1 displays the obtained results in terms of average gap (among all instances) from Archetti et al. [1] results. That Table also displays the average computing time (among all instances) in seconds (CPU).

Table 1. Comparison between basic ILS heuristics.

	ILS_2-Opt	ILS_2-Opt*	ILS_1-0 Exchange-intra	ILS_1-0 Exchange-Inter	ILS_Or-Opt	ILS_1-1 Exchange-intra	ILS_1-1 Exchange_inter
Gap	5.41	4.94	5.78	5.42	5.76	5.52	5.07
CPU	15.70	14.98	15.72	16.42	14.58	15.15	15.92

The results of Table 1 show that the 2-Opt* operator performs better than the other operators in terms of solution quality and computing time.

Hence, in the reminder of the present work, the basic ILS heuristic will refer to the basic ILS heuristic using the 2-Opt* operator. That heuristic will be compared to the other implemented heuristics.

4.3   Study of ILS_LNS Heuristic

Several tests were performed in [29] to determine the best removal/insertion couple of operators of LNS. In addition, LNS was compared with ILS_LNS. ILS_LNS was able to reach better quality solutions and was faster than LNS.

In the present Section, we compare the results of ILS_LNS using the selected couple of removal/insertion operators (Source [29]) with those of the basic ILS heuristic presented in Sect. 4.2.

ILS_LNS is run until 50 iterations without improvement are reached. We remarked that, with only 50 iterations, ILS_LNS is more time consuming than the above studied heuristics using 500 iterations. Hence, we maintain ILS_LNS iteration number to 50.

As one can see from Table 2, the average gap of ILS_2-Opt* is slightly better that the average gap of ILS_LNS. Regarding the average computing time, ILS_LNS appears to be slower than ILS_2-Opt*.

Table 2. Comparison between ILS_LNS and ILS_2-Opt*.

	ILS_LNS	ILS_2-Opt*
Gap	5.07	4.94
CPU	36.35	14.98

4.4   Study of ILS_RVND Heuristic

As the basic ILS, ILS_RVND is very fast in comparison with the ILS_LNS heuristic. Hence, we decide to fix the number of iterations without improvement in the ILS embedded in ILS_RVND to 500. Results of ILS_RVND using 50 iterations are provided in [29].

Table 3 compares the results of ILS_RVND (Source [29]) with both ILS_LNS (Source [29]) and ILS_2-Opt*. We remark that the average gaps of ILS_2-Opt* and ILS_RVND are quite similar. In addition, the average computing time of ILS_RVND is slightly better than the average computing time of ILS_2-Opt*. We think that difference between the two heuristics can be more evident if the iteration number increases and/or if the heuristics are hybridized with other ones.

Table 3. Comparison between ILS_LNS, ILS_2-Opt* and ILS_RVND.

	ILS_LNS	ILS_2-Opt*	ILS_RVND
Gap	5.07	4.94	4.95
CPU	36.35	14.98	10.57

4.5 Study of LNS_RVND Heuristic

LNS_RVND is described in Sect. 3.3. The results related to this hybrid heuristic are presented in [29].

We remark that the results of LNS_RVND (Source [29]) are quite "disappointing" in comparison with other approaches. Indeed, the heuristic obtains the worst average gap and computing time. That can be seen in Table 4.

From the results displayed in Table 4, we conclude that the multi-start ILS heuristic has a considerable impact on the solution quality and the speed of finding solutions.

Table 4. Comparison between ILS_LNS, ILS_2-Opt*, ILS_RVND and LNS_RVND.

	ILS_LNS	ILS_2-Opt*	ILS_RVND	LNS_RVND
Gap	5.07	4.94	4.95	11.94
CPU	36.35	14.98	10.57	67.10

4.6 Study of ILS_2-Opt*_LNS Heuristic

ILS_2-Opt*_LNS is described in Sect. 3.3. In order to have a good balance between solution quality and computing time, the 2-Opt*_LNS heuristic is repeated 7 times at each iteration of ILS. The number of iterations without improvement of the embedded ILS heuristic $maxOcc$ is set to 200. While the number of iterations without improvement of LNS $maxOcc_{LNS}$ is fixed to 20.

Table 5 presents the results of ILS_2-Opt*_LNS compared with those of the previously studied heuristics. We remark that ILS_2-Opt*_LNS is able to reach the best average gap in comparison with the other heuristics. ILS_2-Opt*_LNS seems to be relatively time consuming. However, the computing time of this heuristic is still reasonable, especially if we take the solution quality into account.

Table 5. Comparison between ILS_2-Opt*_LNS and the previously studied heuristics.

	ILS_LNS	ILS_2-Opt*	ILS_RVND	LNS_RVND	ILS_2-Opt*_LNS
Gap	5.07	4.94	4.95	11.94	1.72
CPU	36.35	14.98	10.57	67.10	36.64

4.7 Study of ILS_RVND_LNS Heuristic

ILS_RVND_LNS is described in Sect. 3.3. As in ILS_2-Opt*_LNS, the local search phase of ILS_RVND_LNS, which consists in the RVND_LNS heuristic, is repeated 7 times at each iteration of ILS. The numbers of iterations without improvement in the ILS heuristic $maxOcc$ and in the LNS heuristic $maxOcc_{LNS}$ are set to 200 and 20 respectively.

Note that, contrary to the LNS_RVND studied in Sect. 3.3, the combination of LNS and RVND involved in ILS_RVND_LNS uses the RVND heuristic with a probability equal to 1. That modification is motivated by the fact that ILS_RVND gives good quality solutions in comparison with both ILS_LNS and LNS_RVND.

Table 6 compares the results of ILS_RVND_LNS (Source [29]) with those of the other approaches presented in the present work. From that Table, we remark that ILS_RVND_LNS provides the best average gap without being too time consuming. When comparing ILS_RVND_LNS with ILS_2-Opt*_LNS, we can see that ILS_RVND_LNS is also better in terms of average results (gaps and computing time). That confims our assumption that the use of RVND instead of a neighborhood operator in the ILS heuristic can lead to better results when using more iterations and/or when ILS is hybridized with other heuristics (more than two heuristics are involved).

We also remark that better average gaps are obtained when the heuristics are hybridized. However, that hybridizations can lead to an increase of the computing time. We think that the hybridized heuristics are more time consuming because they are first trapped in local optima then they extract themselves from these optima. That process is repeated several times. On the other hand, the basic heuristic (with a basic level of hybridization or no hybridization at all) are quickly trapped in local optima.

Table 6. Comparison between ILS_RVND_LNS and the other proposed heuristics.

	ILS_LNS	ILS_2-Opt*	ILS_RVND	LNS_RVND	ILS_2-Opt*_LNS	ILS_RVND_LNS
Gap	5.07	4.94	4.95	11.94	1.72	1.57
CPU	36.35	14.98	10.57	67.10	36.64	28.39

4.8 Study of the Perturbation Procedure

In all the studied approaches described so far, we use the *random removal* and the *basic greedy* heuristic (both described in [32]).

Initially, we wanted to use the *random removal* combined to a random insertion heuristic in order to change the characteristics of the solution after a local search is performed. We thought that the two random heuristics can lead to a more diversified search and thus, to better results.

In practice, we found that the use of the *random removal* combined to a random insertion introduces too much diversification. That diversification could not be correctly handled by the local search procedure. Hence, we decided to use the *random removal* combined with the *basic greedy* heuristic instead.

Figure 2 describes the gaps obtained by both versions of the perturbation procedure: with *basic greedy* and the random insertion. The tests are performed using ILS_RVND_LNS.

In Fig. 2, the gap for each instance is shown. From that Figure, we can see that the perturbation using the *basic greedy* generally provides better gaps. Actually the random insertion outperforms *basic greedy* in only 4 cases. The average computing time reached when using *basic greedy* is slightly worse that the average computing time reached when using the random insertion. Indeed, the farmer is equal to 37.85 seconds while the latter is equal to 29.52 seconds.

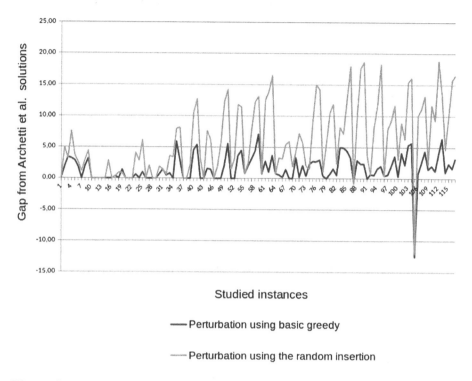

Studied instances

—— Perturbation using basic greedy

···· Perturbation using the random insertion

Fig. 2. Comparison between the use of the basic greedy operator and the random insertion for the perturbation procedure.

4.9   Comparison of ILS_RVND_LNS with Other Approaches from the Literature

ILS_RVND_LNS gives the best results among all the implemented approaches. Therefore, it is compared with other approaches from the literature.

ILS_RVND_LNS is compared with TF, TA and VNS heuristics proposed by Archetti et al. [1].

In order to have relatively comparable results for the compared approaches (in terms of solution quality and computing time), we execute ILS_RVND_LNS 20 times at each run.

Table 7 provides the detailed results. In that Table, *ins* stands for the instance name, where an instance $pXY - m - Q$ refers to the original instance pXY of the CVRP in which the vehicle number is fixed to m and the capacity bound is set to Q. n stands for the customer number. VNS, TF, TA and ILS_RVND_LNS refer to the objective values obtained by VNS, TF, TA and ILS_RVND_LNS respectively. CPU is the computing time of ILS_RVND_LNS is seconds. $CPU(min)$ stands for the average computing time in minutes of all the heuristics among all the instances.

Note that the detailed computing time of VNS, TF and TA can be found in [1].

From Table 7, we remark that ILS_RVND_LNS provides 6 new best solutions. In addition, the heuristic reaches the literature solutions in 55 cases. ILS_RVND_LNS is not able to reach the literature solutions in 56 cases. However, the average gap of ILS_RVND_LNS is relatively small. Indeed, it is equal to 0.66%.

The computing time of ILS_RVND_LNS is reasonable.

Table 7. Detailed results for CPTP instances.

Ins	n	VNS	TF	TA	ILS_RVND_LNS	CPU
p06-10-160	50	258,97	258,97	255,38	**259,12**	356,73
p07-20-140	75	**534,81**	525,06	527,90	524,39	1006,98
p08-15-200	100	**663,98**	657,31	656,32	649,67	2330,63
p09-10-200	150	1189,33	**1192,68**	1143,65	1162,55	4194,31
p10-20-200	199	**1773,65**	1761,37	1759,81	1741,43	10736,14
p13-15-200	120	284,71	269,74	274,28	**289,59**	2417,94
p14-10-200	100	**890,44**	886,78	888,18	**890,44**	1283,71
p15-15-200	150	**1168,63**	1156,01	1134,17	1157,38	4945,23
p16-20-200	199	**1791,78**	1764,15	1776,41	1747,06	8435,07
p06-2-50	50	**33,88**	**33,88**	**33,88**	**33,88**	38,71
p07-2-50	75	**49,18**	**49,18**	**49,18**	**49,18**	51,73
p08-2-50	100	**57,75**	**57,75**	**57,75**	**57,75**	76,92
p09-2-50	150	**65,03**	63,89	**65,03**	**65,03**	97,6
p10-2-50	199	**70,87**	**70,87**	**70,87**	**70,87**	129,57
p13-2-50	120	**64,12**	**64,12**	**64,12**	**64,12**	129,66
p14-2-50	100	**43,26**	**43,26**	**43,26**	**43,26**	76,34
p15-2-50	150	**64,98**	**64,98**	**64,98**	**64,98**	107,39
p16-2-50	199	**66,81**	**66,81**	**66,81**	66,39	142,77
p06-3-50	50	**40,95**	**40,95**	**40,95**	**40,95**	56,78
p07-3-50	75	**69,94**	**69,94**	**69,94**	**69,94**	74,12
p08-3-50	100	**80,82**	**80,82**	**80,82**	**80,82**	121,37

Table 7. (*continued*)

p09-3-50	150	**96,16**	**96,16**	**96,16**	**96,16**	162,38
p10-3-50	199	**103,79**	**103,79**	**103,79**	**103,79**	209,51
p13-3-50	120	**87,25**	**87,25**	**87,25**	**87,25**	207,5
p14-3-50	100	**59,43**	**59,43**	**59,43**	**59,43**	74,87
p15-3-50	150	**96,42**	**96,42**	**96,42**	**96,42**	173,42
p16-3-50	199	**99,7**	**99,7**	**99,7**	**99,7**	228,97
p06-4-50	50	**45,43**	**45,43**	**45,43**	**45,43**	77,98
p07-4-50	75	**90,65**	**90,65**	**90,65**	**90,65**	99,74
p08-4-50	100	**100,36**	98,47	**100,36**	99,76	174,71
p09-4-50	150	**121,35**	**121,35**	**121,35**	**121,35**	247,87
p10-4-50	199	**134,81**	**134,81**	**134,81**	**134,81**	297,66
p13-4-50	120	**104,18**	103,73	103,72	103,34	243,39
p14-4-50	100	**68,63**	**68,63**	**68,63**	**68,63**	133,17
p15-4-50	150	**124,02**	**124,02**	**124,02**	119,52	252,23
p16-4-50	199	**131,37**	**131,37**	**131,37**	**131,37**	327,39
p06-2-75	50	**72,28**	**72,28**	**72,28**	**72,28**	49,4
p07-2-75	75	**92,44**	**92,44**	**92,44**	**92,44**	69,34
p08-2-75	100	**106,15**	**106,15**	**106,15**	**106,15**	143,26
p09-2-75	150	**117,66**	**117,66**	**117,66**	**117,66**	136,25
p10-2-75	199	**124,85**	**124,85**	**124,85**	**124,85**	167,55
p13-2-75	120	**110,12**	**110,12**	**110,12**	**110,12**	155,25
p14-2-75	100	**77,09**	**77,09**	**77,09**	**77,09**	70,38
p15-2-75	150	**120,93**	**120,93**	**120,93**	**120,93**	144,84
p16-2-75	199	**123,38**	**123,38**	**123,38**	**123,38**	179,58
p06-3-75	50	**92,32**	**92,32**	**92,32**	**92,32**	75,7
p07-3-75	75	**131,12**	**131,12**	**131,12**	**131,12**	106,59
p08-3-75	100	**147,55**	**147,55**	145,87	**147,55**	238,69
p09-3-75	150	**160,96**	**160,96**	**160,96**	160,66	238,63
p10-3-75	199	**177,9**	**177,9**	176,50	176,22	263,4
p13-3-75	120	**139,37**	137,95	137,45	**139,37**	375,34
p14-3-75	100	**112,56**	112,51	**112,56**	**112,56**	104,9
p15-3-75	150	**174,58**	**174,58**	**174,58**	**174,58**	225,78
p16-3-75	199	**179,55**	**179,55**	179,23	177,35	279,69
p06-4-75	50	**99,37**	**99,37**	**99,37**	**99,37**	100,6
p07-4-75	75	**158,11**	**158,11**	**158,11**	**158,11**	147,84
p08-4-75	100	**185,27**	**185,27**	**185,27**	181,42	400,73
p09-4-75	150	**204,25**	203,24	203,24	201,47	335,87
p10-4-75	199	**229,27**	**229,27**	**229,27**	225,22	381,49

Table 7. (*continued*)

p13-4-75	120	**161,62**	160,68	157,98	161,59	548,71
p14-4-75	100	**139,88**	139,67	139,83	**139,88**	187,11
p15-4-75	150	**219,22**	**219,22**	216,61	**219,22**	318,75
p16-4-75	199	**235,03**	**235,03**	**235,03**	228,49	392,83
p06-2-100	50	**100,27**	99,50	99,50	**100,27**	63,84
p07-2-100	75	**132,7**	**132,7**	**132,7**	**132,7**	103,65
p08-2-100	100	**158,21**	**158,21**	**158,21**	**158,21**	134,63
p09-2-100	150	**161,23**	**161,23**	**161,23**	161,15	176,16
p10-2-100	199	**171,24**	**171,24**	**171,24**	171,19	189,8
p13-2-100	120	**145,75**	145,67	145,67	**145,75**	248,41
p14-2-100	100	**125,29**	**125,29**	**125,29**	**125,29**	123,27
p15-2-100	150	**169,71**	**169,71**	**169,71**	**169,71**	181,41
p16-2-100	199	**177,23**	**177,23**	175,57	173,56	217,8
p06-3-100	50	**134,72**	**134,72**	**134,72**	**134,72**	93,74
p07-3-100	75	**185,25**	184,88	**185,25**	184,88	167,88
p08-3-100	100	**218,63**	**218,63**	218,33	218,43	284,91
p09-3-100	150	**230,49**	229,61	229,61	229,58	284,75
p10-3-100	199	**250,18**	246,56	246,95	246,56	316,77
p13-3-100	120	**181,63**	177,76	180,04	180,79	625,97
p14-3-100	100	**182,31**	179,48	**182,31**	**182,31**	195,82
p15-3-100	150	**244,08**	241,84	**244,08**	243,89	279,08
p16-3-100	199	**258,07**	257,10	252,44	255,38	333,9
p06-4-100	50	**153,3**	**153,3**	152,97	152,97	120,51
p07-4-100	75	**233,4**	**233,4**	232,05	226,61	225,8
p08-4-100	100	**268,34**	266,23	266,08	259,2	504,64
p09-4-100	150	**290,54**	**290,54**	290,15	285,3	433,15
p10-4-100	199	**324,02**	321,17	321,03	320,07	508,78
p13-4-100	120	200,62	178,82	183,66	**202,21**	1208,47
p14-4-100	100	**237,68**	236,50	**237,68**	**237,68**	251,91
p15-4-100	150	**308,07**	305,30	304,81	302,78	437,42
p16-4-100	199	**336,24**	328,20	329,53	328,29	515,7
p06-2-9	50	**168,6**	**168,6**	**168,6**	**168,6**	93,97
p07-2-9	75	**199,97**	**199,97**	**199,97**	**199,97**	131,79
p08-2-9	100	**330,14**	319,28	319,28	328,37	242,88
p09-2-9	150	**347,9**	347,43	**347,9**	343,72	299,39
p10-2-9	199	**382,41**	378,32	379,81	376,35	362,3
p13-2-9	120	**239,57**	238,58	230,59	238,58	1225,12
p14-2-9	100	**303,17**	302,94	**303,17**	303,14	201,63

Table 7. (*continued*)

p15-2-9	150	**378,09**	**378,09**	**378,09**	376,16	322
p16-2-9	199	**394,05**	390,47	391,71	389,21	372,33
p06-3-9	50	**219,36**	218,96	218,96	218,67	153,06
p07-3-9	75	**274,8**	**274,8**	**274,8**	273,27	207,18
p08-3-9	100	**447,15**	444,82	433,38	444,87	538,03
p09-3-9	150	**500,17**	496,84	500,12	488,79	521,54
p10-3-9	199	**559,8**	549,83	551,44	533,15	608,92
p13-3-9	120	250,69	234,99	244,96	**283,15**	1955,02
p14-3-9	100	418,28	416,32	417,32	**419,63**	331,84
p15-3-9	150	**519,39**	517,18	512,83	513,09	534,16
p16-3-9	199	**567,24**	558,61	558,10	556,53	608,02
p06-4-9	50	**258,97**	**258,97**	254,47	**258,97**	213,45
p07-4-9	75	**344,35**	343,12	339,95	342,7	303,67
p08-4-9	100	536,64	**537,66**	536,13	535,26	902,31
p09-4-9	150	**639,72**	635,67	633,64	621,47	785,67
p10-4-9	199	**723,47**	710,59	719,13	684,68	904,46
p13-4-9	120	279,43	264,46	294,46	**295,77**	2016,09
p14-4-9	100	**537,24**	516,20	531,53	531,94	444,02
p15-4-9	150	653,22	**654,94**	652,58	651,14	803,79
p16-4-9	199	729,40	**731,14**	726,22	719,14	888,78
CPU(min)		10,3	2,83	8,54	9,94	

5   Conclusion

In the present work, we propose a set of basic and hybrid heuristic approaches based on the ILS heuristic. The basic heuristics combine ILS with seven neighborhood operators which results in seven basic ILS heuristics. The hybrid ILS heuristics use either LNS or RVND, or a combination of LNS and a neighborhood operator, or a combination of LNS and RVND. A simple heuristic using only LNS and RVND is also provided to highlight the importance of the ILS heuristic. In addition, two perturbation procedures are tested.

The proposed approaches are evaluated on a variant of the *Vehicle Routing Problem* called *Capacitated Profitable Tour Problem*. The obtained results show that the more ILS is hybridized the better are the results.

The best implemented approach in term of average results is compared with other approaches from the literature. The experimentations show that the proposed heuristic is able to provide competitive results for the *Capacitated Profitable Tour Problem*.

A future work may consists in evaluating the performance of the proposed hybrid heuristic on other variants of the *Vehicle Routing Problem*.

References

1. Archetti, C., Feillet, D., Hertz, A., Speranza, M.G.: The capacitated team orienteering and profitable tour problems. J. Oper. Res. Soc. **60**, 831–842 (2009)
2. Lourenço, H.R., Martin, O.C., Stützle, T.: Iterated local search. In: Glover, F., Kochenberger, G.A. (eds) Handbook of Metaheuristics, vol. 57. Springer, Boston (2003). https://doi.org/10.1007/0-306-48056-5_11
3. Martins, A.X., Duhamel, C., Mahey, P., Saldanha, R.R., de Souza, M.C.: Variable neighborhood descent with iterated local search for routing and wavelength assignment. Comput. Oper. Res. **39**, 2133–2141 (2012)
4. Martins, I.C., Pinheiro, R.G., Protti, F., Ochi, L.S.: A hybrid iterated local search and variable neighborhood descent heuristic applied to the cell formation problem. Expert Syst. Appl. **42**, 8947–8955 (2015)
5. Chen, P., Huang, H.k., Dong, X.Y.: Iterated variable neighborhood descent algorithm for the capacitated vehicle routing problem. Expert Syst. Appl. **37**, 1620–1627 (2010)
6. Subramanian, A., Drummond, L.M.D.A., Bentes, C., Ochi, L.S., Farias, R.: A parallel heuristic for the vehicle routing problem with simultaneous pickup and delivery. Comput. Oper. Res. **37**, 1899–1911 (2010)
7. Subramanian, A., Uchoa, E., Ochi, L.S.: A hybrid algorithm for a class of vehicle routing problems. Comput. Oper. Res. **40**, 2519–2531 (2013)
8. Assis, L.P., Maravilha, A.L., Vivas, A., Campelo, F., Ramírez, J.A.: Multiobjective vehicle routing problem with fixed delivery and optional collections. Optimization Letters **7**, 1419–1431 (2013)
9. Subramanian, A., Battarra, M.: An iterated local search algorithm for the travelling salesman problem with pickups and deliveries. J. Oper. Res. Soc. **64**, 402–409 (2013)
10. Hernández-Pérez, H., Rodríguez-Martín, I., Salazar-González, J.J.: A hybrid heuristic approach for the multi-commodity pickup-and-delivery traveling salesman problem. Eur. J. Oper. Res. **251**, 44–52 (2016)
11. Todosijević, R., Hanafi, S., Urošević, D., Jarboui, B., Gendron, B.: A general variable neighborhood search for the swap-body vehicle routing problem. Comput. Oper. Res. **78**, 468–479 (2017)
12. Erdoğan, G., Cordeau, J.F., Laporte, G.: The pickup and delivery traveling salesman problem with first-in-first-out loading. Comput. Oper. Res. **36**, 1800–1808 (2009)
13. Hernández-Pérez, H., Rodríguez-Martín, I., Salazar-González, J.J.: A hybrid grasp/vnd heuristic for the one-commodity pickup-and-delivery traveling salesman problem. Comput. Oper. Res. **36**, 1639–1645 (2009)
14. Rodríguez-Martín, I., Salazar-González, J.J.: A hybrid heuristic approach for the multi-commodity one-to-one pickup-and-delivery traveling salesman problem. J. Heuristics **18**, 849–867 (2012)
15. Gansterer, M., Küçüktepe, M., Hartl, R.F.: The multi-vehicle profitable pickup and delivery problem. OR Spectrum **39**, 303–319 (2017)
16. Sifaleras, A., Konstantaras, I.: Variable neighborhood descent heuristic for solving reverse logistics multi-item dynamic lot-sizing problems. Comput. Oper. Res. **78**, 385–392 (2017)
17. Samà, M., Corman, F., Pacciarelli, D., et al.: A variable neighbourhood search for fast train scheduling and routing during disturbed railway traffic situations. Comput. Oper. Res. **78**, 480–499 (2017)

18. Hassannayebi, E., Zegordi, S.H.: Variable and adaptive neighbourhood search algorithms for rail rapid transit timetabling problem. Comput. Oper. Res. **78**, 439–453 (2017)
19. Sassi, O., Cherif-Khettaf, W.R., Oulamara, A.: Multi-start iterated local search for the mixed fleet vehicle routing problem with heterogenous electric vehicles. In: Ochoa, G., Chicano, F. (eds.) EvoCOP 2015. LNCS, vol. 9026, pp. 138–149. Springer, Cham (2015). https://doi.org/10.1007/978-3-319-16468-7_12
20. Cuervo, D.P., Goos, P., Sörensen, K., Arráiz, E.: An iterated local search algorithm for the vehicle routing problem with backhauls. Eur. J. Oper. Res. **237**, 454–464 (2014)
21. Silva, M.M., Subramanian, A., Ochi, L.S.: An iterated local search heuristic for the split delivery vehicle routing problem. Comput. Oper. Res. **53**, 234–249 (2015)
22. Morais, V.W., Mateus, G.R., Noronha, T.F.: Iterated local search heuristics for the vehicle routing problem with cross-docking. Expert Syst. Appl. **41**, 7495–7506 (2014)
23. Li, J., Pardalos, P.M., Sun, H., Pei, J., Zhang, Y.: Iterated local search embedded adaptive neighborhood selection approach for the multi-depot vehicle routing problem with simultaneous deliveries and pickups. Expert Syst. Appl. **42**, 3551–3561 (2015)
24. François, V., Arda, Y., Crama, Y., Laporte, G.: Large neighborhood search for multi-trip vehicle routing. Eur. J. Oper. Res. **255**, 422–441 (2016)
25. Grangier, P., Gendreau, M., Lehuédé, F., Rousseau, L.M.: A matheuristic based on large neighborhood search for the vehicle routing problem with cross-docking. Comput. Oper. Res. **84**, 116–126 (2017)
26. Akpinar, S.: Hybrid large neighbourhood search algorithm for capacitated vehicle routing problem. Expert Syst. Appl. **61**, 28–38 (2016)
27. Dominguez, O., Guimarans, D., Juan, A.A., de la Nuez, I.: A biased-randomised large neighbourhood search for the two-dimensional vehicle routing problem with backhauls. Eur. J. Oper. Res. **255**, 442–462 (2016)
28. Canca, D., De-Los-Santos, A., Laporte, G., Mesa, J.A.: An adaptive neighborhood search metaheuristic for the integrated railway rapid transit network design and line planning problem. Comput. Oper. Res. **78**, 1–14 (2017)
29. Chentli, H., Ouafi, R., Cherif-Khettaf, W.R.: Behaviour of a hybrid ils heuristic on the capacitated profitable tour problem. In: Proceedings of the 7th International Conference on Operations Research and Enterprise Systems: ICORES, INSTICC, SciTePress, vol. 1, pp. 115–123 (2018)
30. Solomon, M.M.: Algorithms for the vehicle routing and scheduling problems with time window constraints. Oper. Res. **35**, 254–265 (1987)
31. Talbi, E.G.: Metaheuristics: from design to implementation (2009)
32. Pisinger, D., Ropke, S.: A general heuristic for vehicle routing problems. Comput. Oper. Res. **34**, 2403–2435 (2007)
33. Christofides, N., Mingozzi, A., Toth, P.: The vehicle routing problem. In: Christofides, N., Mingozzi, A., Toth, P., Sandi, C. (eds.) Combinatorial Optimization, pp. 315–338. Wiley, Chichester (1979)

An Efficient Heuristic for Pooled Repair Shop Designs

Hasan Hüseyin Turan[1], Shaligram Pokharel[2(✉)], Tarek Y. ElMekkawy[2],
Andrei Sleptchenko[3], and Maryam Al-Khatib[2]

[1] Capability Systems Centre, School of Engineering and Information Technology,
University of New South Wales, Canberra, Australia
[2] Department of Mechanical and Industrial Engineering, College of Engineering,
Qatar University, Doha, Qatar
`shaligram@qu.edu.qa`
[3] Department of Industrial and Systems Engineering, Khalifa University of Science
and Technology, Abu Dhabi, UAE

Abstract. An effective spare part supply system planning is essential to
achieve a high capital asset availability. We investigate the design problem of a repair shop in a single echelon repairable multi-item spare parts
supply system. The repair shop usually consists of several servers with
different skill sets. Once a failure occurs in the system, the failed part is
queued to be served by a suitable server that has the required skill. We
model the repair shop as a collection of independent sub-systems, where
each sub-system is responsible for repairing certain types of failed parts.
The procedure of partitioning a repair shop into sub-systems is known
as pooling, and the repair shop formed by the union of independent
sub-systems is called a pooled repair shop. Identifying the best partition
is a challenging combinatorial optimization problem. In this direction,
we formulate the problem as a stochastic nonlinear integer programming
model and propose a sequential solution heuristic to find the best-pooled
design by considering inventory allocation and capacity level designation
of the repair shop. We conduct numerical experiments to quantify the
value of the pooled repair shop designs. Our analysis shows that pooled
designs can yield cost reductions by 25% to 45% compared to full flexible and dedicated designs. The proposed heuristic also achieves a lower
average total system cost than that generated by a Genetic Algorithm
(GA)-based solution algorithm.

Keywords: Spare part logistics · Repair shop · Pooling · Heuristic ·
Genetic algorithm

1 Introduction

Service and manufacturing operations rely heavily on the availability of equipment and assets. High availability of assets can be achieved with effective maintenance strategies. However, maintenance can be costly. For example, a recent

© Springer Nature Switzerland AG 2019
G. H. Parlier et al. (Eds.): ICORES 2018, CCIS 966, pp. 102–118, 2019.
https://doi.org/10.1007/978-3-030-16035-7_6

report of IATA's Maintenance Cost Task Force points out that maintenance cost can be anywhere between 10% to 15% of the total operation cost of a commercial airline industry [1]. Similarly, for manufacturing firms, maintenance cost may reach up to 60% of the production cost [2]. Hence, careful planning of maintenance operations not only leads to a decrease in the total cost but also significant improvements in the reliability of systems [3]. Maintenance planning includes the determination of the maintenance strategy (e.g., failure-based/corrective, preventive and condition-based), time interval between maintenance operations, and quantity and quality of maintenance resources such as technicians, supplies and spare parts [4].

In this paper, the corrective maintenance of high-valued assets in and particular the decisions regarding the amount of spare part inventory, capacity and design of repair facilities are investigated. Corrective maintenance of assets is usually done by replacing a failed part by a repaired part available in the stock. If repairable spares are not in the stock, the asset goes down, and a downtime cost is incurred till a sufficient number of spares are supplied to the system [5–7]. A large number of spares are required to ensure a high availability of the capital asset. However, keeping a large number of repairable in inventory increases the cost [8]. The decision on repair shop design heavily influences the number of spares to be stocked. An optimal design of the repair shop can lead to a less number of spare parts that are needed to achieve the same level of availability. Thus, at the operational level, the inventory and repair shop decisions have to be coordinated together to reduce downtimes.

There are different types of repair shop design alternatives, as illustrated in Fig. 1. The two extremes are the full flexible (full cross-training) and the dedicated designs. Figure 1(a) depicts a full cross-training design scheme in which all of the servers are merged into a single cluster/sub-system. In this design scheme, all servers are considered to have necessary skills to repair any type of failed parts. On the contrary, in the dedicated design, each cluster of servers is responsible for repairing a specific type of spare part as in Fig. 1(c).

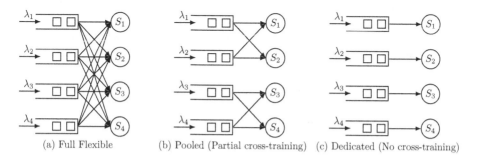

(a) Full Flexible   (b) Pooled (Partial cross-training)   (c) Dedicated (No cross-training)

Fig. 1. Possible repair shop designs.

An intermediate level design scheme between the dedicated and the full flexible system is the pooled design. In the pooled design, the repair shop is a union of independent clusters/sub-systems, where each sub-system is responsible for repairing certain types of failed parts as in Fig. 1(b). In this study, we try to find the best-pooled repair shop design that leads to minimum total system cost. Nevertheless, identifying the best-pooled repair shop design is a challenging combinatorial optimization problem. Thus, we develop an efficient solution heuristic to overcome computational complexity of the problem. The proposed solution heuristic also computes the optimal amount of spare part inventories to keep on stock and the number of servers (capacity) that have to be allocated into each cluster.

The rest of the paper is organized as follows. In Sect. 2, a review of related literature is provided. In Sect. 3, problem definition and the mathematical model are presented. The solution heuristic for the proposed model is discussed in Sect. 4. Section 5 provides a comparative computational study under input settings. Conclusions and future research directions are summarized in Sect. 6.

2   Literature Review

Some significant advances in optimization of spare part supply systems, capacity and inventory theory and design of flexibility service/manufacturing systems can be seen in the literature. However, research opportunities exist at the intersection of these research areas. The optimization problem analyzed in this paper exploits the intersection of the design of a flexible/cross-trained repair shop in a spare part supply system and optimization of resource capacity of the repair shop and inventory levels of spares.

The dominant model for repairable items, both in the literature and in the practical applications, is METRIC (Multi-Echelon Technique for Recoverable Item Control), developed by [5]. METRIC based models assume that the repair capacity is infinite. This assumption may not be appropriate in most industrial settings. Hence, some researchers have relaxed ample repair capacity assumption by explicitly considering finite repair service capacity [9–13]. In addition to the limited repair capacity assumption, integrated optimization of repair capacity and maintenance policies are also studied extensively in the literature by [14–17]. The work of [18] relaxes assumptions of stationary failure rates of spares and finite repair capacity at the same time. Similarly, the model of [19] also optimizes spare inventory levels under finite repair capacity together with nonstationary failure rates of spares under a certain budget restriction or spending over certain availability. The impact of limited but not constant (varying repair capacity on a system for repairable items) is analyzed by [20]. We refer to the recent literature review article of [21] for integrating decisions on spare parts inventories and repair shops.

In [22] regarding the manufacturing resource flexibility, a comparison between a totally dedicated system, a totally flexible system, and several intermediate possibilities are provided. In the following years, several authors have extended the work provided by [22] and some of them have also validated robustness of *"little or limited flexibility"* being usually sufficient for optimal system performance (see Ref. [23–30]). Cross-training is one of the more widely discussed capacity/workforce flexibility methods for complex systems [31]. Production lines [32,33], job shops [34], flow/assembly shops [35–39], manufacturing [40–43], call centers [44–47], health care [48,49], field services [50–53] and maintenance/repair [54] are some examples of systems where cross-training is applied. The more detailed discussions and classifications of flexibility and cross-training applications can be found in the recent review articles of [31,55].

A limited number of cross-training related studies also appears in the maintenance and spare part logistics literature. For example, [50–52] discuss the optimal cross-training policies of technicians/service engineers in a field service setting. Similarly, [53,54] address workforce management problems in corrective repair/maintenance environments, in which repairmen are either cross-trained or dedicated. The analysis of cross-training schemes in repair shop design is discussed in [56–58] by using simulation-optimization techniques.

Even though pooling is considered as a partial cross-training of resources, to the best of our knowledge, no results has been presented analyzing pooling in spare part supply systems other than very recent works of [59–61]. Our work fills the gap on pooled repair shops designs in spare part supply systems integrated with capacity decision in the literature.

3 Problem Description and Formulation

We study the design problem of a repair shop in a single echelon repairable multi-item spare part supply system. The repair shop may consist of several parallel multi-skilled servers, and storage facilities for the repaired items. Once a failed part is received from the technical system at the installed base, it is queued to be served by a suitable server with the required skills. At the same time, if a repaired (as-good-as-new) part is available in the inventory, it is sent back to the installed base. If the item is not available in the stock, the request is backordered. In this case, the technical system goes down and a downtime cost occurs till the requested ready-for-use part is delivered.

The repair shop may have pooled structure with one or more cells/clusters or an arbitrary structure. In arbitrary designs, not all servers in a cluster are fully flexible; i.e., some servers are partially cross-trained to repair only a subset of all stock keeping units (SKUs) in the cluster. In this paper, we restrict design alternatives limited to only pooled repair shops as in Fig. 1(b), and formulate a stochastic mixed-integer mathematical programming model to find the minimum cost spare part supply system.

In this paper, we proceed from commonly used assumptions in a repairable spare part supply system (see ref. [6,56] and assumption lists therein):

(a) The failures of spares occur according to a Poisson process and are mutually independent from each other with constant rates.
(b) The repair times are exponentially distributed and mutually independent. The expected repair times depend on the SKU type and are independent of the processing server.

(c) First come first served (FCFS) queuing discipline is adopted inside each and every cluster, and no priorities exist among the failed spares.
(d) For all parts $(s - 1, s)$ one-for-one replenishment policy is used. That is, the stock level equals s and each demand immediately generates an order for a replacement part; as a consequence, there is no batching.
(e) The total holding costs for every SKU per unit time are linear in the initial inventory levels (initially acquired inventory).
(f) Penalty costs (or backorder costs) occur when the required part is not available and are paid per time unit per not available SKU.
(g) A positive cross-training (or flexibility) cost occurs whenever an additional skill is assigned to a server. In other words, the cross-training cost is an increasing function on the number of skills per server.
(h) Each cluster inside the repair shop is modeled as a multi-class multi-server $M/M/k$ queuing system with dedicated queues; i.e., every server inside a cluster has the ability to repair all SKUs that are assigned to that cluster.
(i) The clusters inside the repair shop are mutually exclusive (disjoint) and collectively exhaustive. That is, a particular failed SKU can be repaired at exactly one cluster and each SKU is assigned to exactly one cluster.

The last two assumptions (h) and (i) restrict the repair shop design alternatives to the pooled designs. These two assumptions also limit the computational complexity of the system and enable using queue-theoretical approximations to find steady-state probability distribution of items in the system.

We use the problem formulation presented in [61]. The sets, parameters and decision variables for the developed formulations and solution procedures are presented as follows.

Decision Variables

S_i: Amount of initial inventory (basestock level) kept on stock for SKU type i $(i = 1, \ldots, N)$, where $\mathbf{S} = (S_1, \ldots, S_N)$.

z_k: Number of the operational servers in the cluster k $(k = 1, \ldots, y)$, and where $\mathbf{Z} = (z_1, \ldots, z_y)$.

x_{ik}: Binary variable indicating that whether the cluster k has a skill to repair SKU type i $(i = 1, \ldots, N)$ or not, where $\mathbf{X}_k = (x_{1k}, \ldots, x_{Nk})^T$ and $\mathbf{X} = [\mathbf{X}_1 | \ldots | \mathbf{X}_y]$.

y: Number of clusters in the repair shop.

Problem Parameters

N: Number of distinct types of repairables (SKUs).
λ_i: Failure rate of SKU type i $(i = 1, \ldots, N)$.
μ_i: Service rate of SKU type i $(i = 1, \ldots, N)$.
h_i: Inventory holding cost of SKU type i per unit time per part $(i = 1, \ldots, N)$.
b: Penalty cost for each back ordered demand per unit time, which is equivalent to paying per unit time per technical system that is down because of a lack of spare parts.

f: Operation cost of a server per unit time (e.g., annual wage).

c_i: Cost of having skills to repair SKU type i per unit time per server $(i = 1, \ldots, N)$ (e.g., annual qualification bonus).

ϵ: Very small positive real number.

The objective function in Eq. (1) has four cost terms namely, server (capacity), cross-training, holding and backorder costs. Objective function considers several trade-offs between the cost terms such as the cost of holding excess inventory and the cost of downtime, and also the trade-off between the cost of having single or several clusters that include dedicated or cross-trained servers.

$$\min_{\mathbf{S}, \mathbf{X}, \mathbf{Z}} \sum_{k=1}^{y} f z_k + \sum_{k=1}^{y} z_k \left(\sum_{i=1}^{N} c_i x_{ik} \right) + \sum_{i=1}^{N} h_i S_i + b \sum_{i=1}^{N} \mathbb{EBO}_i \left[S_i, \mathbf{X}, \mathbf{Z} \right] \quad (1)$$

The *penalty (backorder)* cost term is calculated using the penalty cost b and the expected total number of backordered parts $\mathbb{EBO}_i \left[S_i, \mathbf{X}, \mathbf{Z} \right]$ for each SKU type i in the steady-state; under the given initial inventory level S_i, pooling scheme of the repair shop $\mathbf{X}$ and the server assignment policy $\mathbf{Z}$. The variable $\mathbf{X}$ represents the $(N \times y)$ matrix of the binary decision variables x_{ik} denoting how SKUs are pooled in the repair shop, and the variable $\mathbf{Z}$ represents a $(1 \times y)$ row matrix of integer decision variables z_k denoting the number of servers in each cluster of the repair shop.

Constraints (2) and (4) ensure that pooling scheme $\mathbf{X}$ satisfies mutually exclusive and total exhaustive condition for each cluster, i.e., any SKU type being repaired by exactly one cluster. Queues (number of waiting failed spares) in each cluster have to have finite queue length at the steady-state to prevent overloading of the repair shop. Thus, the stability of the system is guaranteed by constraint (3) and (5) by assigning sufficient number of servers to each cluster. Constraints (4–7) are required for non-negativity and integrality of the variables. For a non-overloaded system, the overall utilization rate of a particular cluster k $(\sum_{i=1}^{N} x_{ik} \lambda_i / \mu_i)$ must be strictly smaller than the capacity (total number of servers in the cluster z_k) of that cluster, which is ensured by the parameter, ϵ.

$$\sum_{k=1}^{y} x_{ik} = 1 \qquad\qquad i = 1, \ldots, N \qquad (2)$$

$$\sum_{i=1}^{N} x_{ik} \frac{\lambda_i}{\mu_i} \leq z_k (1 - \epsilon) \qquad\qquad k = 1, \ldots, y \qquad (3)$$

$$x_{ik} \in \{0, 1\} \qquad\qquad i = 1, \ldots, N \ \ k = 1, \ldots, y \qquad (4)$$

$$z_k \in \mathbb{Z}^+ \qquad\qquad k = 1, \ldots, y \qquad (5)$$

$$S_i \in \mathbb{N}_0 \qquad\qquad i = 1, \ldots, N \qquad (6)$$

$$y \in \{1, \ldots, N\} \qquad\qquad (7)$$

4   Solution Algorithm: Pooling Heuristic

We search for the optimal values of decision variables sequentially by fixing the values of some decision variables and optimizing the remaining ones as discussed in [61]. First, feasible partitions of SKUs, i.e., pooling policies/schemes $\mathbf{X}$ are generated. Pooling schemes are generated either by pooling heuristic or by a genetic algorithm as explained in the Subsect. 5.2. Then, capacity levels $\mathbf{Z}$ and basestock inventory levels $\mathbf{S}$ are optimized under the given pooling scheme for each cluster. The visual flow of the proposed solution heuristic(s) together with its sub-routines and their interactions with each other are depicted in Fig. 2.

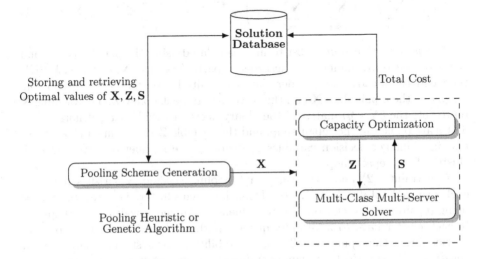

Fig. 2. Flow of the solution algorithms.

In the pooling heuristic, to form partitions of SKUs into clusters, all SKUs are sorted in ascending order by their service rates μ_i so that SKUs closer in service rates are likely to be in the same cluster. This is expected to decrease variations in the service times of SKUs in clusters. Decrease in the variation of service times usually results in a decrease in the number of failed parts waiting for repair in the cluster and eventually lowering the number of backorders and the total cost. Afterward, sorted list of SKUs is divided into smaller lists that have a size of n_{max} or less. The trade-off between the run time of algorithm and the output solution quality are taken into account to determine the value of n_{max}. We set n_{max} as 10 for our experimental runs. For the smaller list ($N \leq n_{max}$), the total enumeration function that is discussed in [61] is invoked. Total enumeration function takes an array of SKU indexes as an input and slices it into sub-arrays for given number of clusters y from 1 to the length of the input array. Each slice/sub-array corresponds to a cluster in a pooled repair shop, and each slic- ing scheme corresponds to a particular pooling policy $\mathbf{X}$. For the larger sorted

SKU index sets, it is not possible to enumerate all slicing schemes with total enumeration function. Therefore, we divide the problem into sub-problems that have the maximum size of n_{max} or less and call total enumeration function for each sub-problem obtained after division. Then, we generate new sub-problems by combining the last and the first elements of adjacent sub-problems. At each iteration, we insert a new SKU index to newly generated sub-problem till the size of the problem reaches n_{max}.

After the generation of the pooling policy $\mathbf{X}$ via above-described pooling heuristic, capacity and inventory level optimization modules are called as shown in Fig. 2. These modules rely on the fact that for every feasible policy $\mathbf{X}$, each cluster can be analyzed and optimized separately due to the clusters being mutually exclusive and independent from each other. The decomposition of the repair shop in sub-systems by pooling reduces the complexity of the problem and enables the use of queue-theoretical approximations to optimize the inventory $\mathbf{S}$ and capacity levels $\mathbf{Z}$. Each cluster k in the repair shop for given number of servers z_k can be analyzed as a multi-class multi-server $M/M/z_k$ queuing system.

The probability distribution of the number of failed SKU type i at the steady-state, $p_i(q)$, is required to evaluate $\mathbb{EBO}_i[S_i, \mathbf{Z}, \mathbf{X}]$ in the objective function. To calculate the probability distribution of the number of failed SKU type i, the approach proposed by [62] is used. Nonetheless, computational burden arises when the number of SKU types and the number of servers increases in the cluster. To overcome this issue, the queuing approximation discussed in [63, 64] is used. In this approximation, marginal probability distribution (and several performance characteristics) of SKU type i in cluster k is derived by aggregating all other SKUs in the cluster k into a single SKU type (class). The procedure is repeated to obtain the remaining distributions for other SKUs in the cluster.

Figure 3 visualizes how N-class $M/M/z_k$ system is decomposed into N independent 3-class $M/M/z_k$ for approximation, where Λ_A and $\Lambda_{A'}$ denote the arrival rates of aggregated classes.

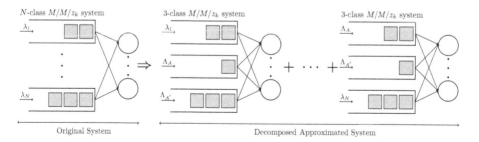

Fig. 3. Approximation of a N-class $M/M/z_k$ queuing system with decomposition into N 3-class $M/M/z_k$ sub-system.

For the given pooling policy $\mathbf{X}$ and capacity levels for each cluster $\mathbf{Z}$, the problem can be reduced to the following one-dimensional optimization problem for each SKU type i by using the independence of each cluster:

$$\min_{S_i \in \mathbb{N}_0} \left(h_i S_i + b \mathbb{EBO}_i \left[S_i, \mathbf{X}, \mathbf{Z} \right] \right) \tag{8}$$

The optimization problem in Eq. (8) takes into account the trade-off between holding and backorder costs, which has similar structure as traditional newsboy problem (see [60] for a detailed discussion). By using the approximated distributions found by queuing approximation $\tilde{p}_i(q)$, the above optimization problem can be minimized by the smallest S_i for which Eq. (9) holds.

$$\sum_{q=0}^{S_i} \tilde{p}_i(q) \geq \frac{b - h_i}{b} \quad i = 1, \ldots, N \tag{9}$$

5   Numerical Study

In this section, we present a computational study of the proposed solution algorithm. First, in Subsect. 5.1, the experiment testbed used in analysis is given and in Subsect. 5.2, details on benchmarking algorithm are provided. In Subsect. 5.3, total system cost reductions achieved by different algorithms are analyzed. Additionally, comparison of cross-training schemes are also given. In Subsect. 5.4, run times of the proposed optimization algorithms are provided.

5.1   Testbed

A full factorial design of experiment (DoE) with 7 factors and 2 levels per factor is used to generate the testbed with total of 128 test instances as in [56,61]. The number of SKUs, N, and the initial total number of servers, M, are the first two DoE factors with levels 10 and 20 for the numbers of SKUs, and 5 and 10 for the initial numbers of servers. The failure rates and the service rates are generated based on the system (repair shop) utilization rate with an assumption that all SKUs are processed on all servers, i.e., a repair shop design with one cluster and fully flexible servers. The system utilization rate, ρ, is the third design factor with levels 0.65 and 0.80. For the chosen utilization rate, we randomly generate two sets of parameters:

(a) the failure rates λ_i, such that $\sum_{i=1}^{N} \lambda_i = 1$, and
(b) workload percentages δ_i, such that $\sum_{i=1}^{N} \delta_i = 1$.

Using the generated λ_i and δ_i, we produce the service rates μ_i as $\mu_i = \frac{\lambda_i}{\delta_i \rho M}$, where $\delta_i \rho M$ is the total workload of SKU type i. The pattern of the holding costs, h_i, is the fourth design factor with two variants (levels): (i) IND: completely randomly (independent) within a range $[h_{min}, h_{max}]$, and (ii) HPB: hyperbolically related to the workloads $w_i = \lambda_i / \mu_i = \delta_i \rho M$:

$$h_i = \frac{h_{max} - h_{min} + 10}{9 \frac{w_i - w_{min}}{w_{max} - w_{min}} + 1} - 10 + h_{min} + \xi_i$$

where

$$\xi_i \in U[-\frac{h_{max} - h_{min}}{20}, \frac{h_{max} - h_{min}}{20}],$$

$$w_{min} = \min_{i=1,...,N} w_i \quad \text{and} \quad w_{max} = \max_{i=1,...,N} w_i$$

The parameters of the hyperbolic relation are chosen such that it replicates some of the real-life scenarios where more expensive repairables are repaired less frequently. The minimum holding cost, h_{min}, is the fifth factor with levels 1 and 100. The maximum holding cost is fixed at 1,000. The server cost, f, and the skill cost, c_i, are the last two factors in our DoE. The server cost levels are set as 10,000 and 100,000 ($10h_{max}$ and $100h_{max}$). The skill cost is assumed as 1% or 10% of the chosen server cost for all SKUs. The penalty cost, b, is set as fifty-fold of the average holding cost so that about 98% of requests can be met from spare stocks. That means the probability of backorder is only 0.02. The overview of all factors and levels are presented in Table 1.

Table 1. Problem parameter variants for test bed [61].

Factors	Levels
No. of SKUs (N)	$[10, 20]$
No. of initial servers (M)	$[5, 10]$
Utilization rate (ρ)	$[0.65, 0.80]$
Minimum holding cost (h_{min})	$[1, 100]$
Maximum holding cost (h_{max})	1000
Holding cost/Workload relation	$[IND, HPB]$
Server cost (f)	$[10h_{max}, 100h_{max}]$
Cross-training cost (c_i)	$[0.01f, 0.10f]$
Penalty cost (b)	$50\frac{\sum_{i=1}^{N} \lambda_i h_i}{\sum_{i=1}^{N} \lambda_i}$

5.2   The Benchmarking Algorithm: A Genetic Algorithm

We compare the performance of the proposed pooling heuristic with a Genetic Algorithm (GA)-based methodology. In this method, a GA searches for the optimal pooled repair shop design policy **X** as it is depicted in Fig. 2.

The GA is a stochastic optimization technique that is inspired by natural selection and biological evolutionary philosophy. A population of individuals (solutions) is represented by a chromosome, a string of information which is randomly generated [57]. Each chromosome corresponds to a particular repair shop design policy, **X**. Every chromosome in the population has N genes. The value of the gene indicates the cluster that SKU is assigned into. Each chromosome also carries information about the number of clusters exist in the repair shop. The total number of distinct integer in the chromosome represents the number of clusters.

At each iteration, GA generates a set of feasible pooled repair shop design policies. Afterward, these candidate feasible solutions (policies) are passed through fitness evaluation function to find optimal values of server assignment policy $\mathbf{Z}$ and inventory levels of spares $\mathbf{S}$. In the fitness evaluation, capacity optimization and multi-class multi-server solver sub-routines are invoked exactly the same way as described for the pooling heuristic. GA runs till it reached predefined generation number. The population size, the number of generations, crossover probability, and mutation probability are the input parameters for any GA implementation. We set the population size and the number of generations at 100 and 25, respectively. Besides, the crossover and the mutation parameters are chosen as 0.8 and 0.4, respectively.

5.3   Performance Comparison of Pooling Heuristic and GA

We find the optimal pooled designs together with optimal capacity and inventory levels of spares for the cases described above by using the proposed pooling heuristic. We compare the minimum total system cost achieved by pooling heuristic with the cost obtained from GA-based pooling algorithm. We define a cost-ratio metric Δ, a ratio of the total minimum cost obtained from the pooling heuristic to the total minimum cost achieved by GA-based algorithm.

Table 2 presents average values of Δ under each problem factor and level. First, on an average, pooling heuristic achieves around 3% lower total cost than that of GA-based pooling algorithm. Second, the increasing size of the problem (higher number of SKUs, N) leads to more substantial objective function value gaps in favor of the pooling heuristic. It shows that the proposed pooling heuristic

Table 2. Total cost comparison of different solution algorithms under varying factors.

Factor	Levels	Cost-ratio Δ	# of the best cost	
			GA-based pooling	Pooling heuristic
Number of SKUs (N)	10	1.0009	17	47
	20	0.9331	1	63
Number of initial servers (M)	5	0.9629	10	54
	10	0.9711	8	56
Utilization rate (ρ)	0.65	0.9646	7	57
	0.80	0.9694	11	53
Minimum holding cost (h_{min})	1	0.9649	9	55
	100	0.9691	9	55
Holding cost/Work load relation	IND	0.9695	8	56
	HPB	0.9645	10	54
Server cost (f)	$10h_{max}$	0.9563	5	59
	$100h_{max}$	0.9776	13	51
Cross-training cost (c_i)	$0.01f$	0.9600	5	59
	$0.1f$	0.9740	13	51
Overall		0.9670	18	110

conducts a more extensive search in a larger solution space (i.e., higher number of SKUs, N). Lastly, the pooling heuristic outperforms the GA-based pooling optimization in 86% of the cases (110 cases out of 128) in the testbed.

We also compare the total system cost with the costs obtained from fully flexible (a single cluster where any SKU can be processed on any server) and dedicated (where the number of clusters equal to the number of SKUs) designs. Table 3 summarizes the cost reduction for both pooling heuristic and GA-based pooling optimization under different problem factors. The repair shop designs found by pooling heuristic can produce approximately 45% and 25% savings on average in comparison with dedicated and fully flexible designs, respectively. In some extreme settings, average cost reduction achieved by pooling heuristic reaches to 55% to that of a dedicated design and 40% to that of a fully flexible design. The repair shop designs suggested by GA-based pooling bring about 44% and 21% total cost reductions compared with dedicated and fully flexible designs, respectively. When the cost of having an extra skill is relatively high compared to that of having an additional server (i.e., the case of cross-training cost being equal to $0.1f$), fully flexible design becomes as good as dedicated design. However, when cross-training cost is relatively small, both of the solution algorithms exhibit worse performance with respect to fully flexible design.

Table 3. Average cost reductions in comparison with dedicated and fully flexible systems.

Factor	Levels	GA-based pooling		Pooling heuristic	
		Dedicated	Fully Flexible	Dedicated	Fully Flexible
Number of SKUs (N)	10	35.93%	21.89%	35.19%	22.00%
	20	52.40%	21.18%	55.65%	28.11%
Number of initial servers (M)	5	52.62%	19.37%	53.76%	23.43%
	10	35.71%	23.70%	37.08%	26.68%
Utilization rate (ρ)	0.65	46.73%	23.36%	48.01%	24.30%
	0.80	41.60%	19.71%	42.83%	25.81%
Minimum holding cost (h_{min})	1	44.59%	21.58%	45.84%	24.81%
	100	43.75%	21.49%	45.00%	25.30%
Holding cost/Work load relation	IND	43.94%	21.29%	44.42%	24.04%
	HPB	44.39%	21.79%	46.42%	26.08%
Server cost (f)	$10h_{max}$	38.63%	17.19%	39.95%	22.44%
	$100h_{max}$	49.71%	25.89%	50.89%	27.67%
Cross-training cost (c_i)	$0.01f$	48.81%	9.54%	50.22%	9.93%
	$0.1f$	39.53%	33.53%	40.62%	40.18%
Average		**44.16%**	**21.53%**	**45.42%**	**25.06%**

Figure 4 shows distributions of the average percentage of cross-training per server for all 128 instances investigated in this paper. We observe that, in most of the instances, the average percentage of cross-training is less than 40%, which shows that partial flexibility; i.e., partial cross-training is usually sufficient for optimal system performance.

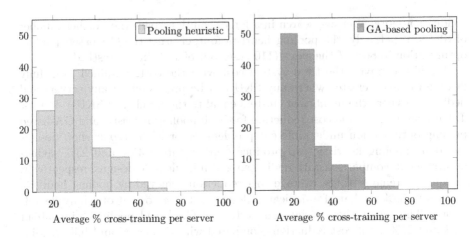

Fig. 4. Cross-training analysis of different solution algorithms.

5.4 Run Time Comparisons

All the experiments are implemented on a computer with 16 GB RAM and 2.8 GHz i7 CPU. Figure 5 shows boxplots of run time performances for both algorithms. The pooling heuristic converges quite fast in most of the cases and provides the final solution within 5000 cpu seconds with a median run time of 2000 s. GA-based pooling algorithm outperforms the pooling heuristic in terms of run times by achieving under 1000 s median run time. Even the worst run time performances of the algorithms are still acceptable for tactical and operational level decisions in real-life spare part supply systems.

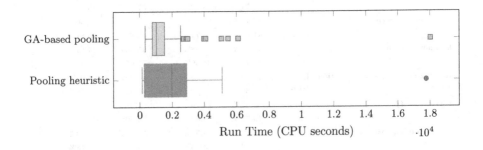

Fig. 5. Run time performance comparison of algorithms.

6   Conclusions

When designing a spare part supply network for repairable parts that balances cost efficiency with effectiveness, several questions in both strategic and tactical nature have to be answered. In this article, the joint problem of resource pooling, inventory allocation and capacity level designation of the repair shop is analyzed, and solution heuristics are developed and compared with each other. From the numerical experiments, it can be concluded that pooled designs result in cost savings of around 45% and 25% in comparison to dedicated and fully flexible designs, respectively. Besides, we observe that the optimal repair shop designs can be achieved by partially cross-trained servers.

The results of this research are important to maintenance outsourcing companies and large firms that operate and maintain their own repair facilities. In both cases, the goal of decreasing maintenance costs and reducing the production stoppages and losses would be accomplished.

As further research possibilities, testing the applicability of the methodology with real-life cases (with larger problem sizes; i.e, a larger number of SKUs) would be an invaluable contribution. We plan to develop novel clustering heuristics or meta-heuristics that generate better pooling schemes with less computational complexity. It would be also worthwhile to integrate pooling decision with static and dynamic routing and prioritization rules in the part repair processes.

Acknowledgement. This research was made possible by the NPRP award [NPRP 7-308-2-128] from the Qatar National Research Fund (a member of The Qatar Foundation). The statements made herein are solely the responsibility of the author[s].

References

1. IATA's Maintenance Cost Task Force: Airline maintenance cost: executive commentary (2015). https://www.iata.org/whatwedo/workgroups/Documents/ MCTF/AMC-Exec-Comment-FY14.pdf. Accessed 30 Aug 2017
2. Keizer, M.C.O., Teunter, R.H., Veldman, J.: Clustering condition-based maintenance for systems with redundancy and economic dependencies. Eur. J. Oper. Res. **251**(2), 531–540 (2016)
3. López-Santana, E., Akhavan-Tabatabaei, R., Dieulle, L., Labadie, N., Medaglia, A.L.: On the combined maintenance and routing optimization problem. Reliab. Eng. Syst. Saf. **145**, 199–214 (2016)
4. Duffuaa, S.O.: Mathematical models in maintenance planning and scheduling. In: Ben-Daya, M., Duffuaa, S.O., Raouf, A. (eds.) Maintenance, Modeling and Optimization, pp. 39–53. Springer, Boston (2000). https://doi.org/10.1007/978-1-4615-4329-9_2
5. Sherbrooke, C.C.: Metric: a multi-echelon technique for recoverable item control. Oper. Res. **16**(1), 122–141 (1968)
6. Sherbrooke, C.C.: Optimal Inventory Modeling of Systems: Multi-echelon Techniques, vol. 72. Springer, New York (2004). https://doi.org/10.1007/b109856
7. Basten, R., Van Houtum, G.: System-oriented inventory models for spare parts. Surv. Oper. Res. Manag. Sci. **19**(1), 34–55 (2014)
8. Arts, J.: A multi-item approach to repairable stocking and expediting in a fluctuating demand environment. Eur. J. Oper. Res. **256**(1), 102–115 (2017)
9. Diaz, A., Fu, M.C.: Models for multi-echelon repairable item inventory systems with limited repair capacity. Eur. J. Oper. Res. **97**(3), 480–492 (1997)
10. Rappold, J.A., Van Roo, B.D.: Designing multi-echelon service parts networks with finite repair capacity. Eur. J. Oper. Res. **199**(3), 781–792 (2009)

11. Sleptchenko, A., Van der Heijden, M., Van Harten, A.: Trade-off between inventory and repair capacity in spare part networks. J. Oper. Res. Soc. **54**(3), 263–272 (2003)

12. Srivathsan, S., Viswanathan, S.: A queueing-based optimization model for planning inventory of repaired components in a service center. Comput. Ind. Eng. **106**, 373–385 (2017)

13. Sleptchenko, A., Van der Heijden, M., Van Harten, A.: Effects of finite repair capacity in multi-echelon, multi-indenture service part supply systems. Int. J. Prod. Econ. **79**(3), 209–230 (2002)

14. de Smidt-Destombes, K.S., van der Heijden, M.C., van Harten, A.: Joint optimisation of spare part inventory, maintenance frequency and repair capacity for k-out-of-n systems. Int. J. Prod. Econ. **118**(1), 260–268 (2009)

15. de Smidt-Destombes, K.S., van der Heijden, M.C., van Harten, A.: Availability of k-out-of-n systems under block replacement sharing limited spares and repair capacity. Int. J. Prod. Econ. **107**(2), 404–421 (2007)

16. de Smidt-Destombes, K.S., van der Heijden, M.C., Van Harten, A.: On the interaction between maintenance, spare part inventories and repair capacity for a k-out-of-n system with wear-out. Eur. J. Oper. Res. **174**(1), 182–200 (2006)

17. de Smidt-Destombes, K.S., van der Heijden, M.C., van Harten, A.: On the availability of a k-out-of-n system given limited spares and repair capacity under a condition based maintenance strategy. Reliab. Eng. Syst. Saf. **83**(3), 287–300 (2004)

18. Lau, H.C., Song, H.: Multi-echelon repairable item inventory system with limited repair capacity under nonstationary demands. Int. J. Inventory Res. **1**(1), 67–92 (2008)

19. Yoon, H., Jung, S., Lee, S.: The effect analysis of multi-echelon inventory models considering demand rate uncertainty and limited maintenance capacity. Int. J. Oper. Res. **24**(1), 38–58 (2015)

20. Tracht, K., Funke, L., Schneider, D.: Varying repair capacity in a repairable item system. Procedia CIRP **17**, 446–450 (2014)

21. Driessen, M.A., Rustenburg, J.W., van Houtum, G.J., Wiers, V.C.S.: Connecting inventory and repair shop control for repairable items. In: Zijm, H., Klumpp, M., Clausen, U., Hompel, M. (eds.) Logistics and Supply Chain Innovation, pp. 199–221. Springer, Cham (2016). https://doi.org/10.1007/978-3-319-22288-2_12

22. Jordan, W.C., Graves, S.C.: Principles on the benefits of manufacturing process flexibility. Manage. Sci. **41**(4), 577–594 (1995)

23. Jordan, W.C., Inman, R.R., Blumenfeld, D.E.: Chained cross-training of workers for robust performance. IIE Trans. **36**(10), 953–967 (2004)

24. Bassamboo, A., Randhawa, R.S., Mieghem, J.A.V.: A little flexibility is all you need: on the asymptotic value of flexible capacity in parallel queuing systems. Oper. Res. **60**(6), 1423–1435 (2012)

25. Brusco, M.J., Johns, T.R.: Staffing a multiskilled workforce with varying levels of productivity: an analysis of cross-training policies*. Decis. Sci. **29**(2), 499–515 (1998)

26. Brusco, M.J.: An exact algorithm for a workforce allocation problem with application to an analysis of cross-training policies. IIE Trans. **40**(5), 495–508 (2008)

27. Chou, M.C., Chua, G.A., Teo, C.P., Zheng, H.: Design for process flexibility: efficiency of the long chain and sparse structure. Oper. Res. **58**(1), 43–58 (2010)

28. Pinker, E.J., Shumsky, R.A.: The efficiency-quality trade-off of cross-trained workers. Manuf. Serv. Oper. Manag. **2**(1), 32–48 (2000)

29. Tsitsiklis, J.N., Xu, K., et al.: On the power of (even a little) resource pooling. Stoch. Syst. **2**(1), 1–66 (2012)

30. Andradóttir, S., Ayhan, H., Down, D.G.: Design principles for flexible systems. Prod. Oper. Manag. **22**(5), 1144–1156 (2013)
31. Qin, R., Nembhard, D.A., Barnes II, W.L.: Workforce flexibility in operations management. Surv. Oper. Res. Manag. Sci. **20**(1), 19–33 (2015)
32. Hopp, W.J., Tekin, E., Van Oyen, M.P.: Benefits of skill chaining in serial production lines with cross-trained workers. Manage. Sci. **50**(1), 83–98 (2004)
33. Liu, C., Yang, N., Li, W., Lian, J., Evans, S., Yin, Y.: Training and assignment of multi-skilled workers for implementing seru production systems. Int. J. Adv. Manuf. Technol. **69**(5–8), 937–959 (2013)
34. Sayın, S., Karabatı, S.: Assigning cross-trained workers to departments: a two-stage optimization model to maximize utility and skill improvement. Eur. J. Oper. Res. **176**(3), 1643–1658 (2007)
35. Hopp, W.J., Oyen, M.P.: Agile workforce evaluation: a framework for cross-training and coordination. IIE Trans. **36**(10), 919–940 (2004)
36. Li, Q., Gong, J., Fung, R.Y., Tang, J.: Multi-objective optimal cross-training configuration models for an assembly cell using non-dominated sorting genetic algorithm-II. Int. J. Comput. Integr. Manuf. **25**(11), 981–995 (2012)
37. Inman, R.R., Jordan, W.C., Blumenfeld, D.E.: Chained cross-training of assembly line workers. Int. J. Prod. Res. **42**(10), 1899–1910 (2004)
38. Tiwari, M., Roy, D.: Application of an evolutionary fuzzy system for the estimation of workforce deployment and cross-training in an assembly environment. Int. J. Prod. Res. **40**(18), 4651–4674 (2002)
39. Vairaktarakis, G., Winch, J.K.: Worker cross-training in paced assembly lines. Manuf. Serv. Oper. Manag. **1**(2), 112–131 (1999)
40. Slomp, J., Bokhorst, J.A., Molleman, E.: Cross-training in a cellular manufacturing environment. Comput. Ind. Eng. **48**(3), 609–624 (2005)
41. Iravani, S.M., Van Oyen, M.P., Sims, K.T.: Structural flexibility: a new perspective on the design of manufacturing and service operations. Manage. Sci. **51**(2), 151–166 (2005)
42. Bokhorst, J.A., Slomp, J., Molleman, E.: Development and evaluation of cross-training policies for manufacturing teams. IIE Trans. **36**(10), 969–984 (2004)
43. Schneider, M., Grahl, J., Francas, D., Vigo, D.: A problem-adjusted genetic algorithm for flexibility design. Int. J. Prod. Econ. **141**(1), 56–65 (2013)
44. Wallace, R.B., Whitt, W.: A staffing algorithm for call centers with skill-based routing. Manuf. Serv. Oper. Manag. **7**(4), 276–294 (2005)
45. Ahghari, M., Balcioglu, B.: Benefits of cross-training in a skill-based routing contact center with priority queues and impatient customers. IIE Trans. **41**(6), 524–536 (2009)
46. Legros, B., Jouini, O., Dallery, Y.: A flexible architecture for call centers with skill-based routing. Int. J. Prod. Econ. **159**, 192–207 (2015)
47. Tekin, E., Hopp, W.J., Van Oyen, M.P.: Pooling strategies for call center agent cross-training. IIE Trans. **41**(6), 546–561 (2009)
48. Harper, P.R., Powell, N., Williams, J.E.: Modelling the size and skill-mix of hospital nursing teams. J. Oper. Res. Soc. **61**(5), 768–779 (2010)
49. Li, L.L.X., King, B.E.: A healthcare staff decision model considering the effects of staff cross-training. Health Care Manag. Sci. **2**(1), 53–61 (1999)
50. Simmons, D.: The effect of non-linear delay costs on workforce mix. J. Oper. Res. Soc. **64**(11), 1622–1629 (2013)
51. Agnihothri, S.R., Mishra, A.K.: Cross-training decisions in field services with three job types and server-job mismatch. Decis. Sci. **35**(2), 239–257 (2004)

52. Agnihothri, S., Mishra, A., Simmons, D.: Workforce cross-training decisions in field service systems with two job types. J. Oper. Res. Soc. **54**, 410–418 (2003)

53. Colen, P., Lambrecht, M.: Cross-training policies in field services. Int. J. Prod. Econ. **138**(1), 76–88 (2012)

54. Iravani, S.M., Krishnamurthy, V.: Workforce agility in repair and maintenance environments. Manuf. Serv. Oper. Manag. **9**(2), 168–184 (2007)

55. De Bruecker, P., Van den Bergh, J., Beliën, J., Demeulemeester, E.: Workforce planning incorporating skills: state of the art. Eur. J. Oper. Res. **243**(1), 1–16 (2015)

56. Sleptchenko, A., Turan, H.H., Pokharel, S., ElMekkawy, T.Y.: Cross training policies for repair shops with spare part inventories. Int. J. Prod. Econ. (2018). https://doi.org/10.1016/j.ijpe.2017.12.018

57. Turan, H.H., Pokharel, S., Sleptchenko, A., ElMekkawy, T.Y.: Integrated optimization for stock levels and cross-training schemes with simulation-based genetic algorithm. In: International Conference on Computational Science and Computational Intelligence, pp. 1158–1163 (2016)

58. Sleptchenko, A., Elmekkawy, T.Y., Turan, H.H., Pokharel, S.: Simulation based particle swarm optimization of cross-training policies in spare parts supply systems. In: The Ninth International Conference on Advanced Computational Intelligence (ICACI 2017), pp. 60–65 (2017)

59. Al-Khatib, M., Turan, H.H., Sleptchenko, A.: Optimal skill assignment with modular architecture in spare parts supply systems. In: 4th International Conference on Industrial Engineering and Applications (ICIEA), pp. 136–140. IEEE (2017)

60. Turan, H.H., Sleptchenko, A., Pokharel, S., ElMekkawy, T.Y.: A clustering-based repair shop design for repairable spare part supply systems. Comput. Ind. Eng. **125**, 232–244 (2018)

61. Turan, H.H., Pokharel, S., Sleptchenko, A., ElMekkawy, T.Y., Al-Khatib, M.: A pooling strategy for flexible repair shop designs. In: Proceedings of the 7th International Conference on Operations Research and Enterprise Systems, pp. 272–278 (2018)

62. Van Harten, A., Sleptchenko, A.: On Markovian multi-class, multi-server queueing. Queueing Syst. **43**(4), 307–328 (2003)

63. Altiok, T.: On the phase-type approximations of general distributions. IIE Trans. **17**(2), 110–116 (1985)

64. Van Der Heijden, M., Van Harten, A., Sleptchenko, A.: Approximations for Markovian multi-class queues with preemptive priorities. Oper. Res. Lett. **32**(3), 273–282 (2004)

A Marginal Allocation Approach to Resource Management for a System of Multiclass Multiserver Queues Using Abandonment and CVaR QoS Measures

Per Enqvist[1] and Göran Svensson[1,2(✉)]

[1] KTH Royal Institute of Technology, Stockholm, Sweden
penqvist@math.kth.se
[2] Teleopti AB, Stockholm, Sweden
goran.svensson@teleopti.com

Abstract. A class of resource allocation problems is considered where some quality of service measure is set against the agent related costs. Three multiobjective minimization problems are posed, one for a system of Erlang-C queues and two for systems of Erlang-A queues.

In the case of the Erlang-C systems we introduce a quality of service measure based on the Conditional Value-at-Risk with waiting time as the loss function. This is a risk coherent measure and is well established in the field of finance. An algebraic proof ensures that this quality of service measure is integer convex in the number of servers.

In the case of the Erlang-A systems we introduce two different quality of service measures. The first is a weighted sum of fractions of abandoning customers and the second is Conditional Value-at-Risk, with the waiting time in queue for a customer conditioned on eventually receiving service. Finally, numerical experiments on the two system types with the given quality of service measures, are presented and the optimal solutions are compared.

Keywords: Queueing · Queueing networks · Marginal allocation · Conditional Value-at-Risk · Abandonments

1 Introduction

This paper is an extended version of [5], which was presented at the ICORES2018 conference in Funchal, Madeira. The text formatting and the sections have been reworked, additional references to previous works has been added and the optimization formulation and the proof of Proposition 1 has been streamlined and a new *Quality of Service* (QoS) measure has been introduced.

In this paper we consider networks of parallel queueing systems of multiclass multiserver type and the server allocation problem where some QoS measure is set against the agent related costs.

© Springer Nature Switzerland AG 2019
G. H. Parlier et al. (Eds.): ICORES 2018, CCIS 966, pp. 119–133, 2019.
https://doi.org/10.1007/978-3-030-16035-7_7

The network consists of queueing systems of one of two types. The first queueing system type is based on the Erlang-C model with *Conditional Value-at-Risk* (CVaR) [14,15] as the QoS measure using the waiting time as the loss function. In the second queueing system type we look at the Erlang-A model, which includes abondonments, for two QoS measures. The measures used are the weighted fraction of abandoning customers and the CVaR with the loss function given by the waiting time of a customer conditioned on eventually receiving service.

The network consists of a system of parallel server pools, see Fig. 1. Each server pool consists of a set of agents (servers) that can serve a specific customer class, and each server pool serves a different customer class. The parallel queueing systems are bound together by one, or more, common budget constraints.

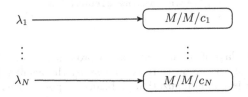

Fig. 1. A system of N parallel $M/M/c$ queues, [5, Fig. 1].

The resource allocation problems described above can be formulated as three multiobjective optimization problems minimizing the different QoS measures and agent cost structures. Using the *Marginal Allocation* (MA) algorithm [6], all efficients solutions can easily be generated. Depending on the actual budget constraints the best solution on the efficient front can be found. The MA algorithm requires that the agent cost structures and the QoS measure are separable and (integer) convex functions.

Following in the footsteps of [4,16,18] we prove that the QoS measure determined by CVaR on the Erlang-C system is decreasing and convex in the number of agents. A recent summary of other convexity proofs for different QoS measures is given in [13].

The main contributions of this paper are the following. The introduction of CVaR as a risk measure for these types of queueing problems, and the benefits of the CVaR measure over VaR type measures is elaborated on. In connection to this we prove that the CVaR measure, using the waiting time as the loss function, is convex and nonincreasing in the number of servers for the Erlang-C type system. We also show, via examples, the similarities between the multiobjective solutions for the Erlang-A and Erlang-C systems.

2   Model Description

We consider a system of $N \in \mathbb{N}$ queues of either $M/M/c$ type (as in Fig. 1) or of $M/M/c + M$ type (using the notation of [3]), i.e., Erlang-C or Erlang-A models. Each server pool has a separate and infinite *first-come-first-serve* (FCFS) buffer.

Introduce the index set $I = \{1, \ldots, N\}$. Let c_i denote the number of servers in queue $i \in I$, and $\mathbf{c} = [c_1 \ \ldots \ c_N]$.

The arrival process to queue $i \in I$ is modelled as a homogeneous Poisson process with arrival rate parameter λ_i, which is often a realistic assumption. The service rate of each server in queue i is denoted by μ_i and the service times are exponentially distributed, which is a tractable model. In the case of the Erlang-A type systems, the abandonment rates of customers is modelled. The time to abandonment, depending on the individual patience, of a customer waiting for service in queue i is exponentially distributed with rate parameter θ_i as in [12]. Hence, the customer may leave the system either due to service completion or impatience, whichever occurs first. It is assumed that customers being served do not defect. The arrival processes, the service times and the individual patience are all independent and the system is in steady state.

The total cost of the agents is assumed to be a separable and increasing integer convex function $g(\mathbf{c}) = \sum_{i=1}^{N} g_i(c_i)$, e.g., a linear cost function $g(\mathbf{c}) = \sum_{i=1}^{N} c_i a_i$ where a_i is a fixed cost for each agent of type $i \in I$. Furthermore, the numbers of agents available for assignment to the different demands may be limited. Let $d_i \in \mathbb{N}$, be the maximum number of available agents of type i, and $\ell_i \in \mathbb{N}$ be a least requirement of agents. Limiting the number of agents can also help avoiding probability issues for the CVaR type measures, see Remark 1.

Let b denote a budget constraint on the system, meaning that the total cost of the assigned servers to the queueing network must be within budget. The model may be extended to include several budget constraints, affecting only a subset of the queues.

3   Quality of Service Measures

To promote a positive customer experience the queueing system is endowed with a QoS measure. Typical QoS measures for contact centers include *average speed of answer* (ASA) and the *telephone service factor* (TSF) [7], also known as *service level* (SL). ASA measures the average time a customer spends waiting on service while TSF considers the *acceptable waiting time* (AWT) that a certain percentile of the customers have to wait before receiving service.

For the QoS measure to be suitable for optimization it is important that it is convex. Many of the relevant QoS measures are convex, and then, for many resource management problems, the "law of diminishing returns" hold. Even if convexity has been shown for many cases, an example with varying buffer size where it does not hold, is demonstrated in [10].

3.1   The Conditional Value-at-Risk Measure

In finance a measure akin to the TSF is used under the name of *Value-at-Risk* (VaR), to quantify the risk of (large) losses for some given loss distribution function. The CVaR measure is often preferred to the VaR type measure due to convexity properties [14,15] and that it is a coherent risk measure [2]. The CVaR measure provide the mean of the extreme outcomes at quantile level $\beta \in [0, 1)$. For $\beta = 0$ it gives the expected value of the given loss function. In general, $\beta > 0$, CVaR offers a means of controlling worst case outcomes, which might be of great importance in fields like healthcare, as pointed out in [13]. It may also be beneficial to use CVaR for any service system were all customers must be handled within a reasonable amount of time.

The general expression for CVaR at quantile level β is

$$\phi(\mathbf{x}) = \frac{1}{1 - \beta} \int\limits_{f(\mathbf{x}, \mathbf{y}) \geq \alpha_\beta(\mathbf{x})} f(\mathbf{x}, \mathbf{y}) p(\mathbf{y}) \, d\mathbf{y}, \tag{1}$$

where $\mathbf{x}$ is the decision vector, $\mathbf{y}$ the uncertainty, p is the density function, f the loss function and α_β the VaR value for quantile β.

We consider two CVaR based QoS measures, one for the Erlang-C system with the customers waiting time in the queue as the loss function, and the measure for the Erlang-A system with loss function given by the waiting time in the queue conditioned on that the customer eventually receive service. We denote the first CVaR measure on the system of Erlang-C queues by CVaR-C and the second on the system of Erlang-A queues by CVaR-A.

First some preliminary results for CVaR-C networks. The waiting time distribution for a $M/M/c$ queue is given by [9]

$$\Pr(W_q > t) = \Pi_q(c, \eta) e^{-(c\mu - \lambda)t} = 1 - F_{W_q}(t), \tag{2}$$

where W_q is the random variable representing the waiting time (in queue) and $\Pi_q(c, \eta)$ is the probability of delay, i.e., of having to wait when there are c homogeneous agents working under load $\eta = \frac{\lambda}{\mu}$. The probability of delay is given by the Erlang-C formula [9],

$$\Pi_q(c, \eta) = \frac{\eta^c / c!}{(1 - \eta/c) \sum\limits_{i=0}^{c-1} \eta^i / i! + \eta^c / c!}. \tag{3}$$

It can be calculated efficiently via the following recursion, similar to [19],

$$\Pi_q(c + 1, \eta) = \frac{\lambda}{\mu} \frac{(c\mu - \lambda) \Pi_q(c, \eta)}{((c + 1)\mu - \lambda) c - \lambda \Pi_q(c, \eta)}. \tag{4}$$

The main advantage of this formula is that it leads to stable numerical calculations. We note that the recursion is initiated by $\Pi_q(1, \eta) = \eta$, and any value larger than one should be interpreted as one, since the larger values corresponds to unstable queues when abandonments are excluded.

The different queues may have different QoS requirements, thus let β_i denote the quantile level for queue $i \in I$ and t_i be the VaR(AWT) value for queue i. The QoS requirement for the VaR type measure is then given by

$$\Pr(W_{q,i} \leq t_i) \geq \beta_i, \quad i \in I, \tag{5}$$

and the corresponding β_i-VaR is defined as

$$t_i(c_i) = \min\{t|F_{W_{q,i}}(t) \geq \beta_i, c_i\} = \frac{\log \Pi_q(c_i, \eta_i) - \log(1 - \beta_i)}{c_i \mu_i - \lambda_i}. \tag{6}$$

Since the distribution $F_{W_{q,i}}(t)$ is known, from (2 and 3), and invertible it is possible to determine this minimum explicitly. We use the notation of t_i to underscore that the VaR value is measured in units of time, in this case.

The corresponding β-CVaR is defined by

$$\phi_{\beta_i}(c_i) = \frac{1}{1 - \beta_i} \int_{t_i(c_i)}^{\infty} t \, dF_{W_{q,i}}(t), \quad i \in I, \tag{7}$$

with explicit formulation for $\Pi_q(c_i, \eta_i) \geq 1 - \beta_i$:

$$\phi_{\beta_i}(c_i) = \frac{1}{c_i \mu_i - \lambda_i} \left(\log \left(\frac{\Pi_q(c_i, \eta_i)}{1 - \beta_i} \right) + 1 \right) = t_i(c_i) + \frac{1}{c_i \mu_i - \lambda_i}, \quad i \in I. \tag{8}$$

where $t_i(c_i)$ is given by (6).

Remark 1 (Probability Atoms and CVaR). *If β is small then a probability atom might have to be handled. To avoid such an atom it is required that $\Pi_q \geq 1 - \beta$ holds. For a complete treatment of CVaR in cases where there is a probability atom see [15]. However, for many realistic choices of parameters this is not an issue and will thus be ignored throughout the rest of the paper.*

The QoS measure given by (6) of the β-CVaR is integer convex and decreasing in the number of agents, c. The use of the MA algorithm rests on this fact. This is formalized in Proposition 1.

Proposition 1 (Integer Convexity of β-CVaR). *Consider a M/M/c queue with constant rate parameters $\mu, \lambda > 0$ such that $c\mu > \lambda$ and that $K < \beta \leq 1$, where K is large enough that there is no probability atom. Then*

$$\phi_{\beta_i}(c_i) = t_i(c_i) + \frac{1}{c_i \mu_i - \lambda_i}, \quad i \in I,$$

is a decreasing and integer convex function in $c_i \in \mathbb{N}$.

The proof is algebra based and is given in the Appendix.

In the case of the CVaR-A type system, the loss function is given by the waiting time in queue for a customer that eventually gets served. This loss function may be easiest to obtain via simulation or approximation. In the examples a simulation based approach is used since this immediately provides the probability density for different quantiles. Simulating the model with c_i servers K times, the waiting times $\{w_q^1, \cdots, w_q^K\}$ are obtained, and the CVaR can be estimated by

$$\Phi_{\beta_i}(c_i) = \frac{1}{K} \frac{1}{1 - \beta} \sum_{k=1}^{K} \left[w_q^k - t_i \right]^+ \tag{9}$$

where t_i, the VaR estimate, is determined by the β quantile of the waiting times.

The CVaR approach is more efficient when the distribution is known, even if a numerical approach has to be used when the distribution function is not invertible, see [14, Theorem 2].

Increasing the θ parameter leads to improvements of the QoS measure for customers that get served, since the waiting time will be reduced due to an increase in the fraction of abandonments. However, since more customers abandon the quality of service will most likely not be perceived as improved. Hence, this QoS measure makes most sense under constant (or limited) abandonment rate, and may be difficult to use for comparisons between systems with different abandonment rates.

According to our numerical experiments we have no reason to believe that this measure would not abide by the law of diminishing returns.

3.2   Probability of Abandonment

Systems of multiserver Erlang-A queues open up for other QoS measures. In [8, 11] different measures for queueing systems with abandonments are considered. Erlang-A queues are stable for arbitrary loads, while Erlang-C queues are only stable when $c\mu > \lambda$. Here we consider systems in steady state, with constant (random) demand. Perhaps the most obvious measure is the probability that a customer will abandon the queue before receiving service. This may occur if the arriving customer finds all servers occupied and thus have to wait in queue. Then the customer will abandon if his/her patience runs out before a server becomes available. Let the Erlang-A system with a QoS measure based on the fraction of abandonments be denoted by Pr-A.

Let $\pi_j^{(i)}$ denote the probability of queue $i \in I$ being in state $j \in \mathbb{N}$, where the states are given by the number of customers in queue i. Furthermore, let $E_{1,c_i}^{(i)}$ denote the Erlang blocking formula of the i:th queue with c_i servers. The incomplete Gamma function is defined as

$$\gamma(x, y) = \int_0^y t^{x-1} e^{-t} dt. \tag{10}$$

Then, in accordance with [11], let

$$A(x, y) = \frac{x e^y}{y^x} \gamma(x, y), \quad x > 0, y \geq 0. \tag{11}$$

The probability of an arrival finding all servers busy is given by

$$\Pr(W_q > 0) = \frac{A(\frac{c\mu}{\theta}, \frac{\lambda}{\theta}) E_{1,c}}{1 + \left(A(\frac{c\mu}{\theta}, \frac{\lambda}{\theta}) - 1\right) E_{1,c}}. \tag{12}$$

The fraction of customers abandoning, conditioned on having to wait on arrival, is given by

$$\Pr(Ab|W_q > 0) = \frac{1}{\rho A(\frac{c\mu}{\theta}, \frac{\lambda}{\theta})} + 1 - \frac{1}{\rho}, \tag{13}$$

where $\rho = \frac{\lambda}{c\mu}$ is the offered load per agent. Using the definition of conditional probability yields the fraction of customers abandoning the queue:

$$\Pr(\text{Ab}) = \Pr(\text{Ab}|W_q > 0)\Pr(W_q > 0)$$
$$= \left(\frac{1}{\rho A(\frac{c\mu}{\theta}, \frac{\lambda}{\theta})} + \frac{\rho - 1}{\rho}\right)\left(\frac{A(\frac{c\mu}{\theta}, \frac{\lambda}{\theta})E_{1,c}}{1 + \left(A(\frac{c\mu}{\theta}, \frac{\lambda}{\theta}) - 1\right)E_{1,c}}\right). \qquad (14)$$

The QoS measure given by the fraction of customers abandoning is decreasing and proven to be integer convex in the number of servers for $\mu \geq \theta$ [1]. According to numerical tests we have performed, which agree with the findings of [1,10], we are convinced it is convex for general θ.

4   Optimization Formulation

We pose a general optimization framework based on the queueing systems and the QoS measures defined in Sect. 3.

Consider the multiobjective problem that minimizes the total agent cost and QoS measure for the corresponding queueing systems. The efficient solutions are given by the solutions to the following optimization problem with weighted objective function given in (15) with the weights $\varphi, \psi > 0$,

$$\min_{\mathbf{c}} \quad \varphi\, g(\mathbf{c}) + \psi \sum_{i=1}^{N} \omega(\lambda_i, \mu_i) QoS_i^k(c_i),$$
$$\text{Sub. to } \sum_{i=1}^{N} g_i(c_i) \leq b, \qquad\qquad\qquad (15)$$
$$\ell_i \leq c_i \leq d_i, \qquad\qquad \forall i \in I,$$
$$c_i \in \mathbb{N}, \qquad\qquad \forall i \in I,$$

and where $QoS_i^k(c_i)$ denotes the k:th QoS measure for the i:th queue. The weights φ and ψ determine the relative importance of the two objectives, and the larger ψ is chosen the larger the cost, and the smaller the QoS measure, will be. The weights $\omega(\lambda_i, \mu_i)$ can be used to scale the relative importance of the queueing systems, and thus prioritize queues with higher traffic load or queues where the customers have higher quality demands.

4.1   QoS: CVaR of Waiting Time (No Abandonment)

Consider the optimization problem (15), for the QoS measure defined for CVaR-C discussed in Sect. 3.1, i.e., from (7)

$$QoS_i^1(c_i) = \phi_{\beta_i}^{(i)}(c_i). \qquad\qquad (16)$$

The weights $\omega = 1$ will be used. The lower limits ℓ_i should be chosen so that $\ell_i\mu_i > \lambda_i$ in order to guarantee stability of queue i.

4.2 QoS: Probability of Abandonment

Consider the optimization problem (15) for the QoS measure defined for Pr-A discussed in Sect. 3.2, i.e., from (14)

$$QoS_i^2 = \Pr_i\{Ab|c_i\} \tag{17}$$

For the weights ω we will use the offered load, i.e., $\omega(\lambda_i, \mu_i) = \eta_i$.

Since the $M/M/c+M$ queues are always stable a lower limit ℓ_i is not necessary, but in practice it is recommended to guarantee an acceptable abandonment rate.

4.3 QoS: CVaR for Waiting Time with Abandonment

Consider the optimization problem (15) for the QoS measure defined for CVaR-A discussed in Sect. 3.1, i.e., from (9)

$$QoS_i^3 = \Phi_{\beta_i}^{(i)}(c_i) \tag{18}$$

The weights $\omega = 1$ will be used. The lower limit ℓ_i can be chosen as for Pr-A.

4.4 The Marginal Allocation Algorithm

The MA algorithm is a powerful algorithm for finding the efficient points for two integer convex and separable functions, described in [6,17].

In general when minimizing the multiobjective optimization problem for functions $f, g : \mathbb{N}^N \to \mathbb{R}$ where

$$\begin{aligned}
\Delta f_j(x_j) \leq \Delta f_j(x_j + 1) < 0 \; \forall j, x_j \in \mathbb{N}, \\
0 < \Delta g_j(x_j) \leq \Delta g_j(x_j + 1) \; \forall j, x_j \in \mathbb{N},
\end{aligned} \tag{19}$$

the optimal vector $\mathbf{x}^* \in \mathbb{N}^N$ minimizes $\varphi g(\mathbf{x}) + \psi f(\mathbf{x})$ if and only if the following conditions are satisfied for each $j = 1, \ldots, N$:

$$\begin{cases}
\frac{-\Delta f_j(x_j^*)}{\Delta g_j(x_j^*)} \leq \frac{\varphi}{\psi} \leq \frac{-\Delta f_j(x_j^*-1)}{\Delta g_j(x_j^*-1)} & \text{if } x_j^* > 0, \\
\frac{-\Delta f_j(0)}{\Delta g_j(0)} \leq \frac{\varphi}{\psi} & \text{if } x_j^* = 0.
\end{cases} \tag{20}$$

x_j^* is an efficient solution if and only if there are constants $\varphi, \psi > 0$ such that the conditions (20) are satisfied for each $j = 1, \ldots, N$.

Marginal Allocation Algorithm [17].

Step 0: Generate a table with N columns, fill the columns with the quotients
$-\Delta f_j(n)/\Delta g_j(n)$ for $n = 0, 1, 2, \ldots$
Set $k = 0$, $\mathbf{x}^{(0)} = (0, \ldots, 0)$, $g(\mathbf{x}^{(0)}) = g(\mathbf{0})$ and $f(\mathbf{x}^{(0)}) = f(\mathbf{0})$.
Step 1: Select the largest uncancelled quotient in the table.
Cancel it and let l be the corresponding column number.

Step 2: Let $k := k + 1$ then let $x_l^{(k)} = x_l^{(k-1)}$ and $x_j^{(k)} = x_j^{(k-1)}, \forall j \neq l$.
Let $f(\mathbf{x}^{(k)}) = f(\mathbf{x}^{(k-1)}) + \Delta f_l(x_l^{k-1})$ and $g(\mathbf{x}^{(k)}) = g(\mathbf{x}^{(k-1)}) + \Delta g_l(x_l^{k-1})$.
Terminate algorithm if $g(\mathbf{x}^{(k)}) \geq g^{\max}$; otherwise go to Step 1.

We identify our budget constraint b to be $g^{\max}$ and the constraint on available agents, $d_i, i \in I$, to be the number of quotients to calculate for column i. This algorithm can now be applied to the optimization problem (15) where

$$g(\mathbf{c}) = \sum_{i=1}^{N} a_i c_i \tag{21}$$

with $a_i > 0$, and

$$f_i(c_i) = \begin{cases} QoS_i^1 = \phi_{\beta_i}^{(i)}(c_i) & \text{from (16)}, \\ QoS_i^2 = \Pr_i\{Ab|c_i\} & \text{from (17)}. \\ QoS_i^3 = \Phi_{\beta_i}^{(i)}(c_i) & \text{from (18)}. \end{cases} \tag{22}$$

Given that the costs and the QoS measures in (15) are seperable integer convex functions in the number of servers, the MA algorithm can be used to generate the whole efficient front of the multiobjective optimization problem up to the given upper bound b on the budget.

5   Numerical Examples

The main benefit of using the MA approach is that huge systems can be optimized almost effortlessly. An example of the efficient front for a system of 100 $M/M/c$ queues, with a (maximal) budget constraint of $b = 3000$, is shown in Fig. 2. The input parameters, in terms of arrivals, service rates and agent costs, were uniformly randomly generated with parameters $\lambda_i \in [5, 15]$, $\mu_i \in [0, 2]$, $a_i \in [0, 4]$ and with $\beta = 0.8$. Finding the whole efficient front took less than a second to perform on a laptop.

To compare the Erlang-C and the Erlang-A based systems we look at a three class multiserver system (i.e., three queues). First, find the efficient points for the CVaR-C system, left graph in Fig. 3. Compute the probability of abandonment for these points and compare to the efficient points of the Pr-A system, as depicted in the middle and right graph of Fig. 3 for two different impatience rates. The input parameters are $\boldsymbol{\beta} = [.95 \quad .95 \quad .95]^T$, $\boldsymbol{\lambda} = [15 \quad 10 \quad 20]^T$, $\boldsymbol{\mu} = [0.5 \quad 0.6 \quad 0.7]^T$ and $\mathbf{a} = [12 \quad 15 \quad 18]^T$. The procedure was repeated for two sets of impatience rates, $\boldsymbol{\theta} = [0.25 \quad 0.25 \quad 0.25]^T$ and $\boldsymbol{\theta} = [10 \quad 10 \quad 10]^T$, respectively. The weight function, $\omega(\lambda_i, \mu_i)$, in (17) is chosen as the offered load, η_i.

A similar comparison is made between the solutions of the CVaR-C system and the CVaR-A system. The parameters used were the same as for the previous example except that the patience parameters used were $\boldsymbol{\theta} = [0.25 \quad 0.25 \quad 0.25]^T$ and $\boldsymbol{\theta} = [2 \quad 2 \quad 2]^T$. Since a simulation based approach was used the computational time increased substantially. Simulations work well in an offline situation but suffer in an online environment. The results are shown in Fig. 4.

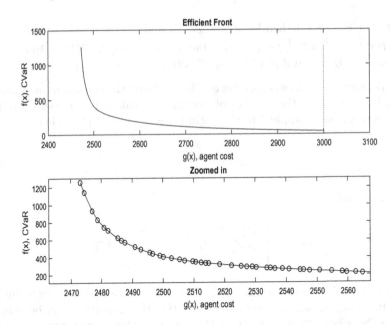

Fig. 2. Showing the efficient front for a system of 100 $M/M/c$-queues, using the measure from CVaR-C as the QoS measure and with a budget constraint of 3000. The parameters where randomly generated. In the bottom figure some specific efficient points are shown.

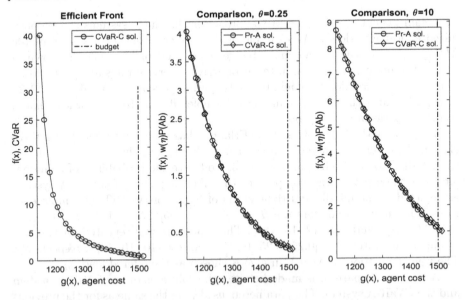

Fig. 3. The left figure shows the CVaR-C solutions for a system of three queues. In the middle and right figures the solutions under abandonments are shown, for patience parameters $\theta = 0.25$ and $\theta = 10$. The Pr-A solutions for abandonments is compared to the CVaR-C solutions in terms of probability of abandonments [5, Fig. 3].

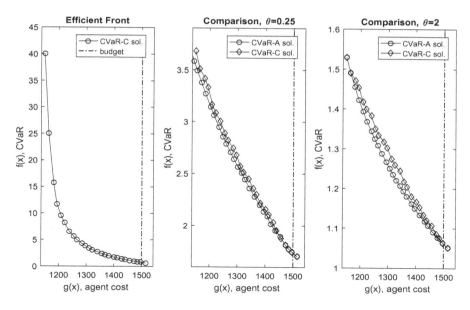

Fig. 4. The left figure shows the CVaR-C solutions for a system of three queues. In the middle and right figures the solutions for CVaR-A is shown, for patience parameters $\theta = 0.25$ and $\theta = 2$. The CVAR-A solutions for the CVAR-A system is compared to the CVaR-C solutions in terms of the CVAR-A QoS measure.

Table 1. Showing the partial agent distributions over the three queues for the CVaR-C solutions (left), for the Pr-A solutions (middle) and CVaR-A solutions (right). The abandonment rate is $\theta = 0.25$ for Pr-A and CVaR-A.

Agents	CVaR-C			Pr-A			CVaR-A		
77	31	17	29	32	17	28	31	19	27
78	31	18	29	33	17	28	32	19	27
79	31	18	30	33	17	29	32	20	27
80	32	18	30	33	18	29	32	21	27
81	32	19	30	34	18	29	32	21	28
82	33	19	30	34	18	30	33	21	28
83	33	19	31	34	19	30	33	21	29
84	33	20	31	35	19	30	33	22	29
85	34	20	31	35	19	31	34	22	29
86	34	20	32	36	19	31	34	23	29
87	35	20	32	36	19	32	35	23	29
88	35	21	32	36	20	32	35	24	29
89	36	21	32	36	20	33	36	24	29
90	36	21	33	37	20	33	37	24	29
91	36	22	33	37	21	33	37	24	30

With respect to the patience parameter, for the example shown in Fig. 3, the two solutions do not differ significantly. The given solutions are very similar, which seem to hold for larger systems and for a wide range of parameter values. A well calibrated CVaR-C measure might be a good alternative to using abandonments, especially since the distribution is invertible and easy to work with. In particular when the load per server is not close to one. The differences between the CVaR-C solution and the CVaR-A solution are more varied. This is to be expected since the load per server for the $M/M/c$-system will naturally be higher.

In both examples the efficient points for the measures on the Erlang-A system might lie close to the solutions generated using the efficient points for the Erlang-C system, but still differ in the distribution of agents. Therefore, we consider the optimal distributions in Table 1 to illustrate this for the examples seen in Figs. 3 and 4. For the Pr-A system the first agent pool is prioritized over the second and third, and receives one or two more agents compared to the CVaR-C assignment. Using CVaR-A it is apparent that the agents in pool 2 are prioritized as compared to the solutions of the CVaR-C system and the opposite holds for pool 3.

6   Summary and Conclusions

The class of resource allocation problems considered in this paper for networks of parallel separable queues are solved using the MA algorithm for three different QoS measures.

Introducing CVaR as a QoS measure for queueing systems allows us to mitigate the worst case service effects. Using the CVaR measure the expected value of the worst outcomes is minimized. Therefore the CVaR measure is well suited for systems where the service of all customers is of importance. For example the TSF does not consider what happens to customers whose service time is longer than the AWT. Another advantage is that the measure is risk coherent.

We provide an algebraic proof of convexity in terms of the number of servers for the CVaR measure for an Erlang-C system with waiting time as the loss function. This fits with "the law of diminishing returns" that holds intuitively for many resource allocation problems. The convexity property is crucial for the application of the MA algorithm, which enables fast solutions.

Examples comparing the performance of solutions determined under different models and QoS measures show that there are variations in the optimal distributions of the servers, but the differences in the resulting QoS measures are relatively minor. The results indicate that the CVaR solution for $M/M/c$ systems may be used to approximate the solutions for systems with abandonment.

Acknowledgements. The authors would like to thank Teleopti AB, Stockholm Sweden, for their support of our work.

Appendix

Proof of Proposition 1

First note that β-CVaR satisfies $\phi_\beta(c, \lambda, \mu) = (1/\lambda)\phi_\beta(c, 1, \mu/\lambda)$, so we can without loss of generality assume that $\lambda = 1$ and $c\mu > 1$.

Let $C_k = (c + k)\mu - 1$, then $C_{k+1} = C_k + \mu$ and $C_k \geq 0$ for all $k \geq 0$. Then, from (7), $\phi_\beta(c) = (\log \Pi_c + 1 - \log(1 - \beta))/C_0$, and the forward difference $\Delta\phi_\beta(c) = \phi_\beta(c + 1) - \phi_\beta(c)$ is given by

$$\Delta\phi_\beta(c) = \frac{1}{C_1}\left[\log\frac{\Pi_{c+1}}{\Pi_c} - \mu\phi_\beta(c)\right] \leq 0 \tag{23}$$

since Π_c is non-increasing, hence $\phi_\beta(c)$ is non-increasing.

The second forward difference is

$$\Delta^2\phi_\beta(c) = \frac{1}{C_1 C_2}\left[2\mu^2\phi_\beta(c) + G\right] \tag{24}$$

and using (4)

$$G = C_1 \log\frac{\Pi_{c+1}}{\Pi_c}\frac{\Pi_{c+1}}{\Pi_{c+2}} - 2\mu\log\frac{\Pi_{c+1}}{\Pi_c} \tag{25}$$

$$= C_1 \log\left(\frac{C_1}{C_0}\frac{C_1 c - \Pi_c}{C_2(c+1) - \Pi_{c+1}}\right) - 2\mu\log\left(\frac{C_0}{(C_1 c - \Pi_c)\mu}\right). \tag{26}$$

Convexity of β-CVaR follows by showing that $G \geq 0$.

Using that $x/(1 + x) \leq \log(1 + x) \leq x$ we have that

$$\log\frac{C_1}{C_2(c+1) - \Pi_{c+1}} \geq -\log(c+1) - \frac{1}{c+1}\frac{\mu(c+1) - \Pi_{c+1}}{C_1}, \tag{27}$$

and

$$\log\frac{C_1 c - \Pi_c}{C_0} \geq \log(c) + \frac{c\mu - \Pi_c}{cC_0 + \mu c - \Pi_c}. \tag{28}$$

Then

$$G \geq C_1 \log\frac{c}{c+1} + 2\mu\log(c\mu) - \mu + \frac{\Pi_{c+1}}{c+1} + \frac{C_3(c\mu - \Pi_c)}{cC_0 + (c\mu - \Pi_c)}. \tag{29}$$

Applying $\log\frac{c}{c+1} \geq -1/c$ and $\log(c\mu) \geq (c\mu - 1)/(c\mu)$ on the first part and (4) on the fourth term, it follows that

$$G \geq \frac{C_{-1}}{c} + \frac{\Pi_c\left(\frac{C_0}{(c+1)\mu} - C_2\right) + 2\mu^2 c}{cC_0 + (c\mu - \Pi_c)}. \tag{30}$$

Note that $cC_0 + c\mu - \Pi_c > 0$ since $c\mu > 1$. Using that $\Pi_c \leq \frac{1}{c\mu+1}\frac{1}{(c-1)\mu+1}$, $G(cC_0 + c\mu - \Pi_c)$ is bounded from below by

$$C_0^2 + \mu^2(2c - 1) - \frac{1}{c\mu + 1}\frac{1}{(c-1)\mu + 1}\left[C_0\left(\frac{c+1}{c} - \frac{1}{(c+1)\mu}\right) + \frac{\mu}{c}(2c - 1)\right] \quad (31)$$

Rearranging, letting $D = (c\mu + 1)((c - 1)\mu + 1)$ then (31) is

$$C_0/D\left[C_0 D - \frac{c+1}{c} + \frac{1}{(c+1)\mu}\right] + \mu\frac{2c-1}{c}\left[c\mu - \frac{1}{D}\right]. \quad (32)$$

First consider the case $c = 1$. Then $\mu > 1$, $C_0 = \mu - 1$, $D = 1 + \mu$ and (32) is given by $B_1(\mu) = 2\mu^3 - 4\mu + 3 + 1/2(1 - 1/\mu)$ which is greater than zero for $\mu > 1$. ($B_1'(\mu) \geq 0$ for $\mu \geq 1$ and $B_1(1) = 1$)

Next consider the case $c \geq 2$ and $\mu \geq 1$. Then $C_0 \geq 1$, $D \geq 2$ and both terms in (32) are greater than zero.

What remains is the case $c \geq 2$ and $\mu < 1$. Then $C_0 \geq 0$, $D \geq 2$. Introduce $\epsilon = \frac{c\mu - 1}{\mu}$, and eliminate c in (31) to obtain

$$\mu^2(\epsilon^2 + 2\epsilon + \frac{2}{\mu} - 1) - \frac{H}{(2 + \mu\epsilon)(2 + \mu\epsilon - \mu)}, \quad (33)$$

where $H = \mu(2 + \epsilon) + \mu^2\frac{\epsilon - 1}{1 + \mu\epsilon} - \frac{1}{1 + \mu + \mu\epsilon}$.

Consider two cases, $\epsilon \geq 1$ and $\epsilon < 1$.

If $\epsilon \geq 1$, then (33) is greater than $\mu^2(\epsilon^2 + 2\epsilon + \frac{2}{\mu} - 1) - H/4$ which, using that $\frac{1}{1+\mu\epsilon} \leq 1$ and $\frac{1}{1+\mu+\mu\epsilon} \geq 1 - \mu - \mu\epsilon$, is greater than

$$(\epsilon\mu + \frac{7\mu - 2}{8})^2 + \frac{12 + 11\mu + 97\mu(1 - \mu)}{64} \quad (34)$$

which is non-negative.

If $\epsilon < 1$, then we write (33) on common denominator and use $\frac{1}{1+\mu\epsilon} \geq 1 - \mu\epsilon$ and $\frac{1}{1+\mu+\mu\epsilon} \geq 1 - \mu - \mu\epsilon$, to show that the numerator is bounded below by

$$(1 - \mu)\left[1 - \epsilon\mu + \mu^3 + 5\mu(1 - \mu) + \mu(1 - \epsilon) + 4\epsilon\mu^2\left(\frac{13}{4} - \mu + \epsilon(1 + \mu)\right)\right]$$
$$+ \mu^3\epsilon^3(4\epsilon + \mu\epsilon^2 + 7) \geq 0 \quad (35)$$

for all $\epsilon, \mu \in (0, 1)$.

Then

$$\Delta^2\phi_\beta(c, 1, \mu) \geq \frac{2\mu^2\phi_\beta(c, 1, \mu)}{C_2 C_1} \geq 0, \quad (36)$$

and ϕ_β is integer-convex, which concludes the proof.

References

1. Armony, M., Plambeck, E., Seshadri, S.: Sensitivity of optimal capacity to customer impatience in an unobservable M/M/S queue (why you shouldn't shout at the DMV). Manufact. Serv. Oper. Manage. **11**(1), 19–32 (2009). https://doi.org/10.1287/msom.1070.0194

2. Artzner, P., Delbaen, F., Eber, J., Heath, D.: Coherent measures of risk. Math. Finan. **9**(3), 203–228 (1999)

3. Baccelli, F., Hebuterne, G.: On queues with impatient customers. Performance **81** (1981)

4. Dyer, M., Proll, L.: On the validity of marginal analysis for allocating servers in "M/M/c" queues. Manage. Sci. **23**(9) (1977)

5. Enqvist, P., Svensson, G.: Multi-server marginal allocation - with CVaR and abandonment based QoS measures. In: Proceedings of the 7th International Conference on Operations Research and Enterprise Systems, ICORES, vol. 1, pp. 297–303. INSTICC, SciTePress (2018). https://doi.org/10.5220/0006652602970303

6. Fox, B.: Discrete optimization via marginal analysis. Manage. Sci. **13**(3), 210–216 (1966)

7. Gans, N., Koole, G., Mandelbaum, A.: Telephone call centers: tutorial, review, and research prospects. Manuf. Serv. Oper. Manage. **5**(2), 79–141 (2003). https://doi.org/10.1287/msom.5.2.79.16071

8. Garnett, O., Mandelbaum, A., Reiman, M.: Designing a call center with impatient customers. Manuf. Serv. Oper. Manage. **4**, 208–227 (2002)

9. Kleinrock, L.: Queueing Systems: Problems and Solutions. Wiley, New York (1996)

10. Koole, G., Pot, A.: A note on profit maximization and monotonicity for inbound call centers. Oper. Res. **59**(5), 1304–1308 (2011). Technical note

11. Mandelbaum, A., Zeltyn, S.: Service engineering in action: the palm/Erlang-A queue, with applications to call centers. In: Spath, D., Fähnrich, K.P. (eds) Advances in Services Innovations, pp. 17–45. Springer, Heidelberg (2007). https://doi.org/10.1007/978-3-540-29860-1_2

12. Palm, C.: Research on telephone traffic carried by full availability groups. Tele **1**, 107 (1957)

13. Parlar, M., Sharafali, M.: Optimal design of multi server markovian queues with polynomial waiting and service costs. Appl. Stoch. Models Bus. Ind. **30**(4), 429–443 (2014)

14. Rockafellar, R., Uryasev, S.: Optimization of conditional value-at-risk. J. Risk **2**, 21–42 (2000)

15. Rockafellar, R., Uryasev, S.: Conditional value-at-risk for general loss distributions. J. Bank. Finan. **26**, 1443–1471 (2002)

16. Rolfe, A.J.: A note on marginal allocation in multiple-server service systems. Manage. Sci. **17**(9), 656–658 (1971)

17. Svanberg, K.: On marginal allocation. Department of Mathematics, KTH, Stockholm (2009)

18. Weber, R.: On the marginal benefit of adding servers to "G/GI/m" queues. Manage. Sci. **26**(9), 946–951 (1980)

19. Zeng, G.: Two common properties of the Erlang-B function, Erlang-C function, and Engset blocking function. Math. Comput. Model. **37**(12), 1287–1296 (2003)

An Efficient Application of Goal Programming to Tackle Multiobjective Problems with Recurring Fitness Landscapes

Rodrigo Lankaites Pinheiro[1,2(✉)], Dario Landa-Silva[2], Wasakorn Laesanklang[3], and Ademir Aparecido Constantino[4]

[1] Webroster Ltd., Peterborough PE1 5NB, UK
rodrigo.pinheiro@webroster.com
[2] ASAP Research Group, School of Computer Science, University of Nottingham, Nottingham, UK
dario.landasilva@nottingham.ac.uk
[3] Department of Mathematics, Faculty of Science, Mahidol University, Nakhon Pathom, Thailand
wasakorn.lae@mahidol.ac.th
[4] Departamento de Informática, Universidade Estadual de Maringá, Maringá, Brazil
aaconstantino@uem.br

Abstract. Many real-world applications require decision-makers to assess the quality of solutions while considering multiple conflicting objectives. Obtaining good approximation sets for highly constrained many-objective problems is often a difficult task even for modern multiobjective algorithms. In some cases, multiple instances of the problem scenario present similarities in their fitness landscapes. That is, there are recurring features in the fitness landscapes when searching for solutions to different problem instances. We propose a methodology to exploit this characteristic by solving one instance of a given problem scenario using computationally expensive multiobjective algorithms to obtain a good approximation set and then using Goal Programming with efficient single-objective algorithms to solve other instances of the same problem scenario. We use three goal-based objective functions and show that on benchmark instances of the multiobjective vehicle routing problem with time windows, the methodology is able to produce good results in short computation time. The methodology allows to combine the effectiveness of state-of-the-art multiobjective algorithms with the efficiency of goal programming to find good compromise solutions in problem scenarios where instances have similar fitness landscapes.

Keywords: Multi-criteria decision making · Goal programming · Pareto optimisation · Multiobjective vehicle routing

© Springer Nature Switzerland AG 2019
G. H. Parlier et al. (Eds.): ICORES 2018, CCIS 966, pp. 134–152, 2019.
https://doi.org/10.1007/978-3-030-16035-7_8

1   Introduction

Tackling highly-constrained optimisation problems with many objectives is difficult even with modern multiobjective algorithms [1]. In real-world scenarios, decision-makers often benefit from having a set of solutions representing a compromise between the multiple objectives so that they can choose the preferred solution(s). It is often useful to use problem domain knowledge during the optimisation in order to obtain better sets of compromise solutions. For example, in the context of continuous multiobjective optimisation problems, [2] estimated Pareto fronts to then obtain values for the decision variables of interesting solutions. Their technique allows to focus the search in sub-regions of the objective space. Another example is the work by [3] using a Bayesian model to learn computationally expensive objective functions to then use the estimation model to explore the search space more quickly.

The multiobjective vehicle routing problem with time windows (MOVRPTW) is a well-know difficult combinatorial optimisation problem that arises in many real-world logistic scenarios [4]. This problem refers to creating a plan for a fleet of identical vehicles to take goods from a depot and deliver them to customers at various locations. Each customer has certain demand level that needs to be satisfied within a specified time window. Objectives usually considered in the MOVRPTW include among others, the minimisation of number of vehicles and the minimisation of total travel distance by all vehicles.

Due to the high number of constraints and objectives in MOVRPTW scenarios, even state-of-the-art multiobjective algorithms struggle to find good approximations to the Pareto optimal front within reasonable computation time. In logistic scenarios where problems like MOVRPTW arise, it is often the case that problem instances corresponding to a different planning periods share parts of the same data. For example, the same or very similar set of vehicles might be available in each planning period. Also, there might be a set of recurring customer orders that need to be satisfied in the different planning periods. This results in the different problem instances presenting recurring features in their fitness landscapes. Other problems like timetabling and personnel scheduling may also have instances with recurring features resulting in similar fitness landscapes (η-dimensional surface representing the Pareto front, where η is the number of objectives).

Previous work proposed a technique to analyse and visualise complex objective relationships and fitness landscapes in multiobjective problems [5,6]. Later, [7] introduced a methodology to exploit the recurring similarity between instances of a multiobjective workforce scheduling and routing optimisation problem, in order to solve instances of the same problem scenario more efficiently. In this methodology, a *pilot* problem instance is solved first using some effective (but not necessarily computationally efficient) multiobjective algorithm to produce an approximation to the Pareto optimal set. Such approximation set is given to the decision-maker so that *target* solutions representing the desired trade-off between the multiple objectives are identified. Then, goal programming is applied with a computationally efficient single-objective solving method, in order to find solutions for other problem instances. In this paper, this methodology is applied to tackle the MOVRPTW in order to further investigate its performance for solving multiobjective problem instances with recurring features. The methodology can be very valuable to facilitate informed decision-making when searching solutions to multiobjective problems. Experiments in this paper are conducted on a set of benchmark instances of the MOVRPTW provided by [8].

Section 2 outlines the multiobjective vehicle routing problem with time windows considered here while Sect. 3 outlines goal programming. Section 4 describes the proposed methodology and Sect. 5 presents the experimental configuration. Sections 6 and 7 present and discuss the results. Section 8 concludes the paper and suggests related future research.

2   Multiobjective Vehicle Routing Problem with Time Windows

A Multiobjective Vehicle Routing Problem with Time Windows (MOVRPTW) is defined on a graph $G = (V, E)$ where V is the set of vertices representing the depot (vertex 0) and the customers (vertices $1 \ldots n$) where each customer has a demand p_i ($i = 1, \ldots, n$). There are h identical vehicles available, each one with capacity Q. In this MOVRPTW, h is considered large enough so that as many vehicles as needed are available to create the routing plan. A set of routes served by the set of vehicles should be created in order to satisfy all demands from all customers. All routes must start and end in vertex 0. The edge set E denotes all possible connections between all vertices. Each edge from vertex i to vertex j has an associated cost, denoted by c_{ij}, that represents distance or time for a vehicle to travel between vertices i and j. Each customer i must be served during their corresponding time window $[a_i, b_i]$. A waiting time is incurred if a vehicle arrives at time $t < a_i$ and hence it must wait until the start of the time window to serve the customer. A delay time is incurred if a vehicle arrives at time $t > a_i$ and hence it must start serving the customer immediately. Once the vehicle starts serving the customer, it stays there for s time until the delivery is completed, this is known as the service time.

[8] proposed a benchmark set of instances for the MOVRPTW with five minimisation objectives: number of vehicles (Z_1), total travel distance by all vehicles (Z_2), makespan or travel time of the longest route (Z_3), total waiting time for all vehicles (Z_4), and total delay time for all vehicles (Z_5). They designed their instances based on different characteristics of the problem and each instance is a combination of these features. The features that constitute a problem instance in these benchmarks are:

- **Number of Customers:** 50, 150 and 250 customers.
- **Time Window:** five different profiles ($tw0, tw1, tw2, tw3, tw4$) of time windows across a planning period of eight hours. These profiles are defined in terms of minutes from the start of the planning period $0 = 8{:}00$ am, $480 = 4{:}00$ pm, etc.). These five time-window profiles are defined as follows:

- *tw*0: [0,480], all customers can be served at any time in the day.
- *tw*1: [0,160], [160,320], [320,480], refers to three types of customers (morning, midday and late).
- *tw*2: [0,130], [175,305], [350,480], also refers to three types of customers as in profile *tw*1 but with shorter time windows.
- *tw*3: [0,100], [190,290], [350,480], also refers to three types of customers as in profile *tw*1 but with longer time windows.
- *tw*4: includes all time-windows from *tw*0, *tw*1, *tw*2 and *tw*3, each customer has one of the 10 time window types in the previous profiles.

- **Demand Types:** three types of demand (10, 20, 30) uniformly distributed.
- **Vehicle Capacity:** the capacity of the vehicles is calculated according to a δ parameter such that $Q = \underline{D} + \delta/100(\overline{D} - \underline{D})$ where $\underline{D}$ is the maximum single demand among all customers and $\overline{D}$ is the sum of all customer demands. The dataset considers three δ values ($\delta 0 = 60$, $\delta 1 = 20$, $\delta 2 = 5$).
- **Service Time:** three values of service time (10, 20, 30) uniformly distributed.

For more details of the MOVRPTW described above and a comprehensive study on the multiobjective nature of the problem, please refer to [9]. There are 45 problem instances and a generator available from https://github.com/psxjpc/MOVRPTW-Generator. The technique to analyse objective relationships described in [6] was applied to these problem instances and results indicate that indeed they have similar fitness landscapes. This is the case even for instances that have different time window profiles, vehicle capacity and the number of customers. However, in this work, we split the 45 problem instances into three datasets according to the number of customers. This decision was taken because even though the fitness landscapes are similar, the scale of the objective values vary considerably according to the number of customers. Therefore, we have 3 datasets each with 15 problem instances, the set VRP-50 with 50 customers, the set VRP-150 with 150 customers and the set of VRP-250 with 25 customers.

3   Goal Programming

Without loss of generality, a multiobjective optimisation problem can be written as minimise $F(x) = (f_1(x), f_2(x), ..., f_n(x))$ subject to $x \in S$, where x is a solution, S is the set of feasible solutions, n is the number of objectives in the problem, $F(x)$ is the image of x in the k-objective space and each $f_i(x)$ is the value of objective i in solution x. For two solutions x and y, it is said that x dominates y, if $\forall i : f_i(x) \leq f_i(y)$ and $\exists j : f_j(x) < f_j(y)$. Moreover, x is said to be *Pareto Optimal* if it is not dominated by any other feasible solution. Then, the aim is to find the set of *Pareto Optimal* solutions usually called *Pareto Set*. This set contains a number of *non-dominated points in the objective space* creating the *Pareto Front*.

Goal programming is one of the earliest proposed approaches to tackle optimisation problems with multiple objective [10]. Basically, goal programming consists of establishing a specific numeric goal for each of the objectives considered in the problem. Then, search is conducted for a solution in which the weighted sum of deviations in the objective values with respect to the goals is minimised.

In order words, goal programming is about establishing a target for each objective and then searching for a solution with objective values as close as possible to those targets. There are three types of goals in goal programming [11]:

- Lower bound: defines a lower value for an objective such that solutions that fall below the lower value are penalised.
- Upper bound: defines an upper value for an objective such that solutions that present higher values than the upper value are penalised. This is the type of goals in the optimisation problem considered here, due to the minimisation nature of all objectives.
- Strict bound: defines a specific target value such that solutions that present values above or below are penalised. This is applicable when obtaining a solution with a specific target value for a given objective is essential. For example, in the case that solutions using exactly h number of vehicles were required in the MOVRPTW.

Once the goals for each objective are set, goal programming techniques derive problem models (LP, MIP, etc.) to find solutions that reach (or are close enough to) the target goals. Several strategies, or goal programming variants, have been presented in the literature. We briefly review the three most widely employed variants [12]:

- **Weighted GP** [13]: used when the decision maker is able to assign an *importance* weight to each goal. The objective function for the problem is then a weighted sum of the deviations from the goals.
- **Lexicographic GP:** when weighting the goals is difficult, but the decision maker is able to prioritise them, the lexicographic GP technique is commonly applied [14]. The deviations to the target goals are minimised according to defined priority levels such that deviations from a higher level goal are considered infinitely more important that deviations from a lower level goal.
- **Chebyshev GP** [15]: consists of minimising the maximum weighted normalised deviation from all the goals, hence promoting solutions that are well-balanced regarding the achievement of the target values.

The weighting and lexicographic methods are considered 'a priory' approaches in the sense that the decision maker should set a ranking between the objectives before conducting the search for solutions. This is not the case in the Chebyshev method which is an 'a posteriori' method because it seeks solutions that are well-balanced in the attainment of all goals so that the decision maker can chose afterwards. In this paper, it is assumed that the decision maker is able to choose a preferred solution from a set of trade-off solutions, instead of being able to establish weights or ranking between the multiple objectives. Hence, only the Chebyshev technique is used later in this work.

A potential issue with goal programming is that it may produce solutions that are not Pareto efficient [16]. This is especially true when the goals are 'pessimistic' and the objectives can be easily achieved. Several methods are proposed to address the issue. Most methods rely on extra information from the decision maker in order to promote the further improvement of certain objectives [17]. Other methods involve extending the search after the solution is found by the goal programming in order to find dominating solutions [18].

Works in the literature usually describe the application of goal programming using exact methods [12,16]. However, many works exist where metaheuristics are employed to solve goal programming models. [19] presents a simulated annealing approach to tackle several test problems of preemptive goal programming. [20] employ a fast converging simulated annealing algorithm to solve a machine-tool selection and operation allocations problem with fuzzy variables. [21] propose a genetic algorithm to tackle a goal programming model for the vehicle routing problem with time windows and [22] presents a genetic algorithm to tackle a goal programming model for a transportation planning problem with three objectives. Goal programming is a sound approach to tackle the MOVRPTW considered here because this technique has been successfully applied to related scheduling and routing problems. For example, it has been applied to nurse scheduling [23,24] and to a version of the vehicle routing problem with soft time-windows [25].

4   The Efficient GP Methodology

Figure 1 shows the overall concept of the methodology which was originally proposed in [7]. Each of the steps is explained below in reference to the MOVRPTW tackled in this paper. The overall idea is to find a set of compromise solutions for a representative instance of the multiobjective problem in hand. The decision maker then selects from this set a solution that exhibits the desirable qualities in respect of the various objectives, without the need to set weights or priorities for the objectives. The objective values in the selected solutions are set as the targets for goal programming when searching for solutions to the other problem instances (e.g. routing plans for other days in the same problem scenario).

1. A *pilot instance* from the dataset with recurring fitness landscape is selected by the decision-maker and solved using multiobjective algorithms to obtain the best possible non-dominated approximation set.
2. The decision-maker chooses a preferred solution t from the obtained non-dominated set. This chosen solution is known as the *target solution* and its objective-vector is denoted by

$$\mathbf{Z}^t = (Z_1^t, Z_2^t, Z_3^t, Z_4^t, Z_5^t)$$

3. Each other instance in the dataset can now be solved with a faster single-objective algorithm using a modified objective function (goal programming variant) aiming to reach the target objective vector.

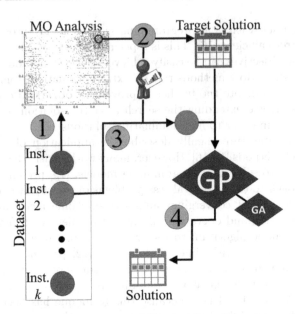

Fig. 1. Overview of the methodology as in [7]. The numbered steps are explained in the main text.

4. The final solution obtained in Step 3 is presented to the decision maker. The overall advantage of this approach is that Step 1, which is typically computationally expensive, needs to be executed only once for a given representative instance in the problem scenario. Then, other problem instances can be solved faster after the target solution is chosen.

The modified objective function of Step 3 has an important role in the methodology as it establishes the way in which the search will aim to attain the goals. Three approaches are used here for determining the objective function. The first one is the well known Chebyshev approach. The second one is to derive a weight-vector from the target solution and the approximation set of the pilot instance. The third approach minimises the Euclidean distances to the target objective-vector.

4.1   Chebyshev Goal Programming

Chebyshev goal programming aims to obtain a balanced solution by minimising the gap to the target of the objective that presents the highest gap, i.e. it seeks to minimise the largest gap to the goals [15]. Hence, if the target goals for the objectives are similarly difficult to attain, this technique can obtain a balanced solution. However, if at least one target objective value is more difficult to achieve (i.e. the target goal is too optimistic), the quality of that objective can be a bottleneck for the other objectives because the search will solely focus on improving that objective. We define the Chebyshev objective function for the MOVRPTW as follows:

$$\text{Minimise} \quad \lambda \tag{1}$$

Subject to

$$\frac{Z_1}{Z_1^t} \leq \lambda \tag{2}$$

$$\frac{Z_2}{Z_2^t} \leq \lambda \tag{3}$$

$$\frac{Z_3}{Z_3^t} \leq \lambda \tag{4}$$

$$\frac{Z_4}{Z_4^t} \leq \lambda \tag{5}$$

$$\frac{Z_5}{Z_5^t} \leq \lambda \tag{6}$$

The Chebyshev objective function given by Eq. (1) is used as the objective function for the MOVRPTW. The main objective is now to minimise λ, thus finding a well-balanced solution regarding reaching the target values. If all targets are reached, λ can assume fractional values and a solution that shows balanced improvements on all objectives may be obtained.

4.2 Derived Weight Vector

One problem with the Chebyshev approach is that it does not guarantee Pareto efficiency. However, the optimal solution for a weighted sum objective function (where weights are not simultaneously null) is always Pareto efficient. To derive a weight vector from the target solution, we first convert the approximation set of the pilot instance into a system of linear inequalities. Considering that the approximation set is composed of k objective-vectors $(\boldsymbol{Z}^1, \boldsymbol{Z}^2, \ldots, \boldsymbol{Z}^5)$, the linear inequalities system can be defined as follows where the aim is to determine the values of $\boldsymbol{w} = (w_1, w_2, w_3, w_4, w_5)$:

$$\begin{cases} \boldsymbol{w}\boldsymbol{Z}^t \leq \boldsymbol{w}\boldsymbol{Z}^1 \\ \boldsymbol{w}\boldsymbol{Z}^t \leq \boldsymbol{w}\boldsymbol{Z}^2 \\ \vdots \\ \boldsymbol{w}\boldsymbol{Z}^t \leq \boldsymbol{w}\boldsymbol{Z}^k \end{cases} \tag{7}$$

There is no guarantee that the system of linear inequalities has a solution if the fitness landscape is non-convex, i.e. if no set of weights can be set to achieve some points in the Pareto optimal front. Therefore, instead of finding a solution for the system, we aim to find a weight vector $\boldsymbol{w}$ that satisfies the largest number of inequalities. Hence, we define the problem of finding the best weight vector as the following MIP (mixed-integer programming) minimisation problem.

$$\text{Minimise} \sum_{j=1}^{k} \overline{x}_j \tag{8}$$

Subject to

$$\boldsymbol{w}\boldsymbol{Z}^t - \boldsymbol{w}\boldsymbol{Z}^j \le M\overline{x}_j \qquad\qquad j = 1, \ldots, k \tag{9}$$

$$w_i \in (0,1], \overline{x}_j \quad \text{binary} \qquad\qquad \begin{cases} i = 1, \ldots, 5 \\ j = 1, \ldots, k \end{cases} \tag{10}$$

The objective function in Eq. (8) aims to find a weight vector $\boldsymbol{w}$ that minimises the number of linear inequalities in (7) which do not fulfill the condition $\boldsymbol{w}\boldsymbol{Z}^t \le \boldsymbol{w}\boldsymbol{Z}^i$ expressed by constraint (9), M is a large constant. Constraint (10) guarantees that zero cannot be chosen as a weight-value (to avoid criteria being removed).

Finally, the weight vector $\boldsymbol{w}$ obtained from the MIP model is used in the objective function for the MOVRPTW as given by Eq. (11).

$$\text{Minimise} \quad \sum_{i=1}^{5} w_i Z_i \tag{11}$$

4.3   Euclidean Distances

We propose an alternative based on the Euclidean distances to the target vector. In essence, this is a method that considers all objectives as equally important. Hence, minimising the Euclidean distances alone does not guarantee Pareto efficiency. In order to mitigate this issue, the proposed method consists of minimising the distances to the target vector for the objectives that are worse than the target. If the current distance for the objectives that are worse than the target vector is small ($<\epsilon$), then the aim is to maximise the distances of the objectives that are better than the target vector.

Henceforth, the objective function in Eq. (12) becomes the objective function for the optimisation problem in hand.

$$\text{Minimize} \begin{cases} z & \text{if } z > \epsilon \\ -z' & \text{otherwise} \end{cases} \tag{12}$$

where

$$z = \sqrt{\sum_{i=1}^{5} z_i} \tag{13}$$

$$z' = \sqrt{\sum_{i=1}^{5} z'_i} \tag{14}$$

$$z_i = \begin{cases} (Z_i - Z_i^t)^2 & \text{if } Z_i > Z_i^t \\ 0 & \text{otherwise} \end{cases} \tag{15}$$

$$z'_i = \begin{cases} (Z_i - Z_i^t)^2 & \text{if } Z_i \leq Z_i^t \\ 0 & \text{otherwise} \end{cases} \tag{16}$$

In summary, when the Euclidean distances of the objectives that are worse than the target vector are larger than the given parameter ϵ, the objective function consists of minimising the Euclidean distances (z). Otherwise, when $z \leq \epsilon$, the objective consists of maximising the distances for the objectives that are better than the target solution (z'). Thus, if the solution has not reached the target, the objective function attempts to close the gap to the target. If the solution is close or better than the target, the objective function attempts to further improve it.

5   Experimental Configuration

We applied the proposed methodology to the MOVRPTW datasets. The instances with $\delta0$ and $tw4$ (50-δ0-tw4, 150-δ0-tw4, 250-δ0-tw4 were arbitrarily selected as pilot instances (Step 1 of methodology). Once the Pareto approximation sets were obtained, $k = 15$ target vectors were randomly selected (uniformly distributed) from each approximation set and the same target vectors were used for the Derived Weight Vector (WV) objective function, the Euclidean Distances (ED) objective function, and the Chebyshev (CV) objective function.

Multiobjective algorithms often struggle to find good approximation sets for combinatorial problems with many objectives (more than three) [1]. Hence, we resort to a tailored procedure to obtain an improved approximation set. [2] state that most multiobjective algorithms can be classified as either Pareto-based or decomposition-based. This study utilises NSGA-II [26] as the Pareto-based algorithm and MOEA/D [27] as the decomposition-based one. Thus, **for each problem instance** the approximation set was obtained (Step 1 of methodology) as described below. The number of solution vectors obtained for each pilot instance was 168 for 50-δ0-tw4, 215 for 50-δ0-tw4 and 206 for 250-δ0-tw4.

1. run both the NSGA-II and MOEA/D for one million objective evaluations on each possible bi-objective vector (Z_1, Z_2), (Z_1, Z_3), ... (Z_4, Z_5);

2. run both the NSGA-II and MOEA/D for one million objective evaluations on each possible three-objective vector (Z_1, Z_2, Z_3), (Z_1, Z_2, Z_4), ... (Z_3, Z_4, Z_5);

3. run both the NSGA-II and MOEA/D for one million objective evaluations on each possible four-objective vector (Z_1, Z_2, Z_3, Z_4), (Z_1, Z_2, Z_3, Z_5), ... (Z_2, Z_3, Z_4, Z_5);

4. create an archive composed of the non-dominated solutions found in the previous three steps;

5. generate a population of individuals where half of the elements are randomly generated and the other half are randomly drawn from the archive built in the previous step;

6. run both the NSGA-II and MOEA/D four times each, for two million objective evaluations, using the initial population generated in the previous step and the five-objective vector; and

7. compile an approximation set with all non-dominated solutions found in all steps.

[28] survey the literature on vehicle routing problem with time windows and show that genetic algorithms are well suited for that problem. Also, our early experiments showed that these algorithms present good enough solutions on these datasets and are simple enough to allow easy replication by other researchers. Hence, for Step 3 of the methodology, the other instances of the MOVRPTW are tackled with a straightforward genetic algorithm (GA) using a direct integer encoding of solutions, uniform crossover, 500 individuals population with a 5% elite being kept across generations and a tournament of two individuals employed for the selection mechanism.

6   Experimental Results

First, we show the effectiveness of the derived weight vector obtained from the MIP model in Eqs. (8)–(10). The effectiveness of a weight vector w is given by the percentage of solutions (in the approximation set for the pilot instance) in which $wZ^t \leq wZ^i$, $i = 1, \ldots, 5$. Hence, if the effectiveness is 100%, it means that the MIP model found a solution for the inequalities system in Eq. (7).

Figure 2 presents the results of the effectiveness analysis. As it was the case in [7] for another problem, the overall effectiveness of the obtained weight vectors here surpassed 90%. Pilot instance 50-δ0-tw4 presented the best average value of 96% and 250-δ0-tw4 presented the worst with 91.3%. Hence, in all cases, the MIP model provided good weight vectors to be used by the WV objective function.

Next, we show the results for each group of instances (for 50, 150 and 250 customers) in three charts. The *target achievement* chart displays the percentage of solutions, in the given dataset, that achieved the target value in each objective. The *gap to target* chart contains the average gap to the target solutions for the solutions that did not reach the target. Finally, the *overall comparison* chart displays the average quality of solutions where positive values indicate that, on

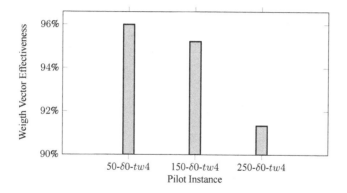

Fig. 2. Average percentage of the solutions in the approximation set of each pilot instance in the MOVRPTW datasets such that $wZ^t \leq wZ^i$.

average, the solutions found are better than the target solution and negative values indicate that the solutions are worse than the target solution.

Figures 3, 4 and 5 display the results of applying the implemented GA with all three objective functions (WV, ED and CV) to the other instances of dataset VRP-50. Results comprise the average values of eight runs for each target vector of each problem instance for each objective function.

Figure 3 shows that for Z_1 and Z_2 the target achievement is close to 100% on all three objective functions. On Z_3 the WV objective function noticeably presents the worst results, with only 63% achievement while the ED and CV objective functions both present similar results with near 80% achievement. Finally, on Z_4 and Z_5 the ED objective function presents a small advantage and the CV objective function is clearly the worst for Z_5.

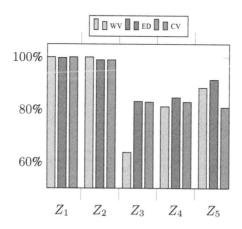

Fig. 3. Dataset VRP-50 – target achievement.

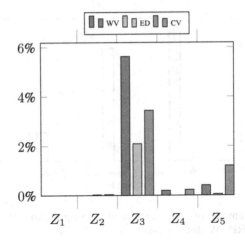

Fig. 4. Dataset VRP-50 – gap to the target.

Figure 4 reflects the findings of the previous figure where Z_3 shown the lowest overall target achievement. Still, on that objective, the overall gap is below 6% for the three objective functions, hence when the target was not met, the gap still was small. Noticeably, the ED objective function presents the lowest gaps. Moreover, Fig. 5 shows that except for WV on Z_3, all objective functions on all objectives present improvements over the target solution, noticeably on Z_1, Z_2 and Z_4 where the solutions found are up to 58% better than the target.

Figures 6, 7 and 8 present the results for the larger set VRP-150. On Fig. 6, we see that while on dataset VRP-50 the objective Z_3 presents the worst results, in this dataset the worst results appear on Z_4 with an average of roughly 75%

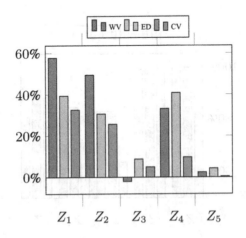

Fig. 5. Dataset VRP-50 – overall comparison.

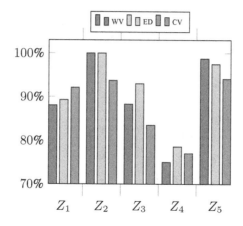

Fig. 6. Dataset VRP-150 – target achievement.

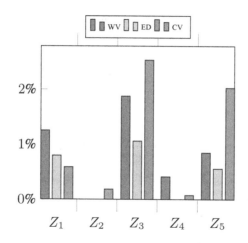

Fig. 7. Dataset VRP-150 – gap to the target.

achievement and, again, the WV objective function presents the worst results. On the other objectives, all objective functions present competitive results.

Figure 7 shows that the gap to the target on solutions that have not met the target is very small – only on Z_2 the gap is larger than 2% and only for the CV objective function.

Figure 8 displays the overall quality of solutions. On average, the quality is better on this dataset than on the previous one. With one or more objective functions, on every objective, the overall quality is more than 20% better than the target. This number increases to nearly 40% for the WV objective function on Z_1 and Z_2.

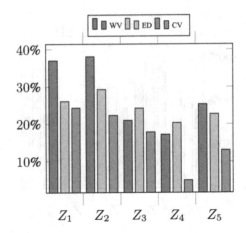

Fig. 8. Dataset VRP-150 – overall comparison.

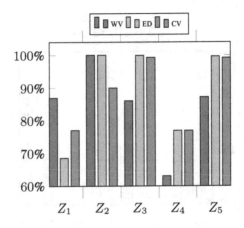

Fig. 9. Dataset VRP-250 – target achievement.

Finally, Figs. 9, 10 and 11 present the results for the largest dataset VRP-250. Figure 9 presents the target achievement. It can be seen that there is a trend, as the size of the datasets increases, the target achievement of Z_1 decreases. In this dataset, the objectives Z_1 and Z_4 presents the worst results. Regarding the objective functions, WV presents the best results for Z_1. On the remaining objectives, the ED objective function presents the most competitive results.

Figure 10 shows the overall gaps to the target solutions. Clearly, the WV approach gets the worst results, even though the gaps were always below 4.2%. Noticeably, the CV objective function presents gaps always smaller than 1%.

Lastly, Fig. 11 shows the overall comparison of solutions with their targets. Again, the results show that all objective functions achieved improved results, with the ED edging Z_3, Z_4 and Z_5 and the WV edging Z_1 and Z_2.

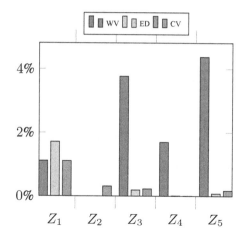

Fig. 10. Dataset VRP-250 – gap to the target.

7   Discussion

While the WSRP datasets tackled in [7] arise from real-world scenarios, the MOVRPTW datasets considered here were fabricated for benchmarking purposes. Also, even the largest MOVRPTW scenario is considerably smaller than a medium-sized WSRP. The target achievement for the MOVRPTW here was larger than in the WSRP overall. The best results obtained here were for the smaller MOVRPTW scenarios, while for the WSRP this happened in the larger instances. We speculate that a reason for this is that the largest MOVRPTW datasets are not large enough for the multiobjective algorithms to struggle in finding good approximation sets (as it happened in the larger WSRP datasets). Therefore, as the performance gap between single-objective algorithms and multiobjective algorithms is considerably smaller in the MOVRPTW problem instances, the difficulties of reaching the target vector becomes more evident.

However, the gaps to the targets of objectives that did not meet their targets were considerably lower here than on the WSRP. Also, the CV objective function, while clearly producing the worst results on the WSRP, it achieves competitive results on the MOVRPTW. This could be a consequence of the quality of the target solutions. The multiobjective algorithms were able to obtain approximation sets with fitness landscape closer the fitness landscape of the optimal Pareto front. Also, there is a higher uniformity of the fitness landscapes across instances for these datasets [6]. All this means that the identified target solutions were realistic, so they could be achieved on every instance. Hence, the CV objective function, which benefits from that, presented good results.

On the MOVRPTW datasets, except for a few exceptions, all objective functions were able to not only reach the target but also to substantially improve all objectives – also a reflection of the quality of the approximation set obtained by the multiobjective algorithms.

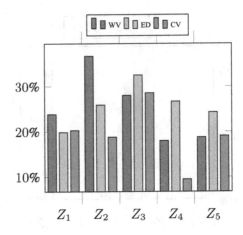

Fig. 11. Dataset VRP-250 – overall comparison.

Nonetheless, it is clear that estimating the Pareto front for problem instances that have similar fitness landscape to the pilot instance, is an effective way to tackle the problem. While the multiobjective algorithms required up to four hours to obtain the approximation set for the pilot instance of a dataset, the GA managed to find competitive solutions in minutes. For the majority of the experiments, targets were achieved and the overall quality of results was high.

8   Conclusion

In this work, we applied a methodology based on goal programming to use efficient single-objective algorithms to solve a multiobjective vehicle routing problem with time windows. The methodology was first presented in [7] and it consists of: (1) solving a pilot instance of the problem using multiobjective algorithms (which are typically computationally expensive) to obtain a good approximation set, (2) having the decision-maker to choose preferred target compromise solutions, and then (3) employing goal programming to solve other instances of the same dataset using the selected solutions in (2) as the target. Three different objective functions were used to guide the search for the target solutions with goal programming. One is the Chebyshev approach that seeks to achieve a solution balanced on all the objective targets. Another one is minimising a weighted function derived from the target solution. The third approach is to use the Euclidean distance to drive the search guided by the target solution.

This methodology was first applied by [7] to real-world instances of a Workforce Scheduling and Routing Problem (WSRP) in the home healthcare sector. In the present paper, the methodology has been further tested by applying it to a different multiobjective problem arising in logistic operational scenarios, the Multiobjective Vehicle Routring Problem with Time Windows (MOVRPTW). In both of these problem scenarios, instances usually arise from different planning periods and hence they present similarities in the fitness landscapes.

This is because usually in this type of real-world problems, instances have the same partial data (e.g. same fleet of vehicles or same set of workers). This paper has shown that the proposed technique is an effective and efficient approach to tackle real-world multiobjective highly-constrained combinatorial optimisation problems, by combining the effectiveness (but often computationally expensive) of state-of-the-art multiobjective algorithms with the efficiency of well-targeted single-objective optimisation through goal programming. For this, the multi-objective analysis technique proposed by [5,6] offers an effective tool to analyse the relationships between objectives in multiobjective optimisation problems and determine the degree of similarity in the fitness landscape of different problem instances.

For future research, it would be interesting to investigate if other approaches besides the Chebyshev, derived weighted function and Euclidean distance approaches, would be more effective across different multiobjective problems. Perhaps an even more interesting but also more challenging future research would be to develop adaptive objective functions that change the search direction as the search progresses and in reaction to the fitness landscape features.

References

1. Giagkiozis, I., Fleming, P.: Methods for many-objective optimization: an analysis. Research Report No. 1030 (2012)
2. Giagkiozis, I., Fleming, P.: Pareto front estimation for decision making. Evol. Comput. **22**, 651–678 (2014)
3. Feliot, P., Bect, J., Vazquez, E.: A Bayesian approach to constrained single- and multi-objective optimization. J. Glob. Optim. **67**(1–2), 97–133 (2017). https://link.springer.com/article/10.1007/s10898-016-0427-3
4. Toth, P., Vigo, D.: The Vehicle Routing Problem. Monographs on Discrete Mathematics and Applications. Society for Industrial and Applied Mathematics (2002)
5. Pinheiro, R.L., Landa-Silva, D., Atkin, J.: Analysis of objectives relationships in multiobjective problems using trade-off region maps. In: Proceedings of the 2015 Annual Conference on Genetic and Evolutionary Computation, GECCO 2015, pp. 735–742. ACM, New York (2015)
6. Pinheiro, R.L., Landa-Silva, D., Atkin, J.: A technique based on trade-off maps to visualise and analyse relationships between objectives in optimisation problems. J. Multi-Criteria Decis. Anal. **24**, 37–56 (2017)
7. Pinheiro, R.L., Landa-Silva, D., Laesanklang, W., Constantino, A.A.: Using goal programming on estimated pareto fronts to solve multiobjective problems. In: Proceedings of the 7th International Conference on Operations Research and Enterprise Systems - Volume 1: ICORES, INSTICC, pp. 132–143. SciTePress (2018)
8. Castro-Gutierrez, J., Landa-Silva, D., Moreno, P.J.: Nature of real-world multi-objective vehicle routing with evolutionary algorithms. In: IEEE International Conference on Systems, Man, and Cybernetics (SMC), pp. 257–264 (2011)
9. Castro-Gutierrez, J., Landa-Silva, D., Moreno Perez, J.: Movrptw dataset (2015). https://github.com/psxjpc/ Accessed 24 Apr 2018
10. Charnes, A., Cooper, W.: Goal programming and multiple objective optimizations. Eur. J. Oper. Res. **1**, 39–54 (1977)
11. Kornbluth, J.: A survey of goal programming. Omega **1**, 193–205 (1973)

12. Jones, D., Tamiz, M.: A review of goal programming. In: Greco, S., Ehrgott, M., Figueira, J. (eds.) Multiple Criteria Decision Analysis. International Series in Operations Research & Management Science, pp. 903–926. Springer, New York (2016). https://doi.org/10.1007/978-1-4939-3094-4_21
13. Romero, C.: Titles of related interest. In: Handbook of Critical Issues in Goal Programming. Pergamon, Amsterdam (1991)
14. Tamiz, M., Jones, D.F., El-Darzi, E.: A review of goal programming and its applications. Ann. Oper. Res. **58**, 39–53 (1995)
15. Flavell, R.B.: A new goal programming formulation. Omega **4**, 731–732 (1976)
16. Jones, D., Tamiz, M.: Goal programming variants. In: Practical Goal Programming, pp. 11–22. Springer, Boston (2010). https://doi.org/10.1007/978-1-4419-5771-9_2
17. Tamiz, M., Mirrazavi, S., Jones, D.: Extensions of pareto efficiency analysis to integer goal programming. Omega **27**, 179–188 (1999)
18. Hannan, E.: Nondominance in goal programming. INFOR Inf. Syst. Oper. Res. **18**, 300–309 (1980)
19. Baykasoglu, A.: Preemptive goal programming using simulated annealing. Eng. Optim. **37**, 49–63 (2005)
20. Mishra, S., Prakash, Tiwari, M.K., Lashkari, R.S.: A fuzzy goal-programming model of machine-tool selection and operation allocation problem in FMS: a quick converging simulated annealing-based approach. Int. J. Prod. Res. **44**, 43–76 (2006)
21. Ghoseiri, K., Ghannadpour, S.F.: Multi-objective vehicle routing problem with time windows using goal programming and genetic algorithm. Appl. Soft Comput. **10**, 1096–1107 (2010)
22. Leung, S.C.H.: A non-linear goal programming model and solution method for the multi-objective trip distribution problem in transportation engineering. Optim. Eng. **8**, 277–298 (2007)
23. Azaiez, M.N., Al Sharif, S.S.: A 0-1 goal programming model for nurse scheduling. Comput. Oper. Res. **32**, 491–507 (2005)
24. Musa, A.A., Saxena, U.: Scheduling nurses using goal-programming techniques. IIE Trans. **16**, 216–221 (1984)
25. Calvete, H.I., Galé, C., Oliveros, M.J., Sánchez-Valverde, B.: A goal programming approach to vehicle routing problems with soft time windows. Eur. J. Oper. Res. **177**, 1720–1733 (2007)
26. Deb, K., Pratap, A., Agarwal, S., Meyarivan, T.: A fast and elitist multiobjective genetic algorithm: NSGA-II. IEEE Trans. Evol. Comput. **6**, 182–197 (2002)
27. Zhang, Q., Li, H.: MOEA/D: a multiobjective evolutionary algorithm based on decomposition. IEEE Trans. Evol. Comput. **11**, 712–731 (2007)
28. Bräysy, O., Gendreau, M.: Vehicle routing problem with time windows, Part II: metaheuristics. Transp. Sci. **39**, 119–139 (2005)

An Efficient Algorithm Based Tabu Search for the Robust Sparse CARP Under Travel Costs Uncertainty

Sara Tfaili[1], Abdelkader Sbihi[1,2(✉)], Adnan Yassine[1,3], and Ibrahima Diarrassouba[1]

[1] Université Le Havre Normandie, LMAH, FR CNRS 3335, ISCN, 76600 Le Havre, France
`sara.tfaili@etu.univ-lehavre.fr`, `abdelkader.sbihi@univ-paris1.fr`, {`adnan.yassine,ibrahima.diarrassouba`}`@univ-lehavre.fr`
[2] École Supérieure de Logistique Industrielle - ESLI, GIP Campus E.S.P.R.I.T. Industries, 35600 Redon, France
[3] Université Le Havre Normandie, ISEL, 76600 Le Havre, France

Abstract. We previously studied the capacitated arc routing problem over sparse underlying graphs under travel costs uncertainty. In this paper, we study the same problem by recalling the mathematical formulation of the problem given in [29]. The problem is characterized by the uncertainty of the travel costs and by the sparse network over which it is defined. In fact, a Multiple-Scenario Min-Max CARP over sparse underlying graphs is studied. More numerical instances applying the greedy heuristic algorithm developed in [29] and the adapted tabu-search algorithm are introduced in which these computational experiments show the effectiveness of these two algorithms.

Keywords: Robust CARP · Travel costs uncertainty · Robust optimization · Scenarios

1 Introduction

The prior knowledge of the data and the parameters of a combinatorial problem are required by the combinatorial optimization. However, perturbations affecting the input data of real life applications leading the data to be unpredictable affect in turn the nature of the solution which will be consequently not optimal or infeasible. For routing problems, this disturbance could be represented by the uncertainty of travel times due to traffic [28], or by the required demands by clients or by the arrivals of new clients for instance. In our work, we study the sparse capacitated arc routing problem with uncertain travel costs by means of a theoretical study and practical one as well.

For the uncertainty in the optimization problems, a contribution to solve the problem has been done using stochastic programming where the uncertain

© Springer Nature Switzerland AG 2019
G. H. Parlier et al. (Eds.): ICORES 2018, CCIS 966, pp. 153–174, 2019.
https://doi.org/10.1007/978-3-030-16035-7_9

data is modeled as random variable with known probability distribution [25, 34]. Applying this type of programming requires two conditions: (1) the stochastic nature of the uncertainties and (2) the possibility of identifying the probability distribution. Heretofore, stochastic programming was the mostly used as in [12] for example, in [9] where a Branch-and-Price algorithm for the capacitated arc routing problem with stochastic demands has been presented and another study is given in [35]. Nevertheless, the nature of the uncertainty in real life problems is not always stochastic where it is not always possible to identify the probability distribution of the data, thus stochastic programming is not always suitable for such problems. As a result, this requires the utilization of an alternative rather than the stochastic programming. In fact, robust optimization that allows determining robust solutions is adaptable for our problem as it provides feasible solutions despite the occurance of any inevitable event or any uncertainty of the input data.

Providing robust solutions addresses three main challenges: (1) evaluation, (2) adaptation of heuristic methods and (3) assessment of performance and this criteria expands to identify (1) the modeling of the uncertain data (scenarios in our study), (2) the selection of appropriate criteria (min-max in our study) and (3) the mathematical model of the problem.

The remaining of the paper is organized as follows. A brief survey about the robust optimization with its definitions and criteria is given in Sect. 2. We introduce a min-max mathematical modeling of our problem which is proposed to give a robust solution minimizing the worst scenario in Sect. 3. Section 4 is consecrated about a heuristic algorithm and an adapted and metaheuristic algorithm which are developed according to specific procedures to determine a solution of the problem and improve it. Computational experiments are performed for the proposed greedy heuristics and metaheuristics in Sect. 5. We derive our conclusions in the last section.

Throughout the whole paper, we work with a graph G whose number of vertices is n, number of edges is $n + \alpha$ with $1 \leq \alpha \leq \frac{n}{2}$, and the maximum vertex degree held in G is 3.

2   Review About Robust Optimization

Robust optimization provides solution that can withstand any uncertainty of the data. Perturbations may affect the data which could affect the optimal solution that is computed before to be infeasible or not optimal. The probabilistic description of the uncertain data allows the use of the stochastic optimization which has been given by Dantzig and Ramser in 1959 [11]. However, this type of programming has two drawbacks:

1. The underlying probability distributions must be already known and this is not always the case.
2. The solutions can become infeasible upon facing some random events or disruptions.

To the contrary, robust optimization which is not stochastic but rather deterministic and set-based is considered to be the suitable alternative where it optimizes the worst case value under all uncertain data. Representing uncertain data is mostly done by a convex set as a polyhedron, a cone or an ellipsoid as seen by Ben-Tal et al. in [2] and by Bertsimas et al. in [5]. Other structures of uncertain data is done by assigning plausible values to each model parameter in which the two common ways of such modeling are the interval or discrete scenarios. In our work, we represent the uncertain data by generating discrete scenarios. In 2010, Sbihi [22], has proposed a robust knapsack based on profits multiple scenario discrete set. In our work, we represent the uncertain data by generating discrete scenarios.

Robustness criteria include several families where the decisions can be made according to max-min, min-max, min-max regret, min-max relative regret and lexicographical min-max, etc. For more details about different robustness criteria, the reader may refer to [10,14,16].

Bertsimas and Sim [6] proposed an alternative robust optimization criterion to the min-max using discrete scenarios. In their paper, they proposed a *budget of uncertainty* denoted by Γ to limit the number of uncertain parameters allowed to deviate from their nominal values, and this proposition is to control the degree of conservatism as robust solutions are often considered as conservative. They applied their method to the cost and constraint coefficients of a linear or mixed integer program which leads to a robust version with a moderate increase in size. They presented a robust integer programming model that allows to control the degree of conservatism of a solution using probabilistic bounds and violation constraints, and they give an algorithm for the robust network flows using the model that they have presented.

Most of the facility location problems test the min-max with discrete scenarios, for example, we may state here the robust prize-collecting Steiner tree problems [1], robust knapsack problem [21] and robust network loading problem with dynamic routing [19]. For more information about robustness criteria and robust classification criteria, see [26].

In our work, the robustness criterion which we follow is the min-max criterion in which we minimize the cost whenever the worst scenario occurs.

2.1   Stochastic Vehicle Routing Problem

A stochastic vehicle routing problem is a VRP where one or several of the components of the problem are random, for instance VRPs with stochastic customers where a customer needs to be serviced with a given probability [3,33], stochastic demands where the demands of the customers are known as probability distributions ([8,18,23,24,27]) and this was deeply studied by Bertsimas [4], or stochastic travel times in which the service or travel times are modeled by random variables (see, e.g., [15,17,32]). A major issue for using the stochastic optimization is that the probability distributions which accurately model uncertainties must be known.

2.2   Robust Vehicle Routing Problem

As mentioned in the previous paragraph, same aspects of uncertainty are handled with the robust vehicle routing problem with a main investigation of that with uncertain demands which was firstly studied by Sungur et al. [28]. They derive a robust counterpart for the vehicle routing problem with stochastic demands (VRPSD) where they show that the robust solution is favorable on average compared to the deterministic solution if the network structure allows a strategical distribution of the slack in the vehicles i.e. in the case where vehicles can share their slacks. Moreover, the authors show that the robust solution is superior to simple strategies of distributing the excess capacity among the vehicles especially when the network structure is more clustered. A limited work has been done for the case of uncertain travel times or travel costs. A robust scenario approach for the vehicle routing problem with uncertain travel times is studied by Han et al. in 2013 [13]. The papers by Toklu et al. [30,31] handle the VRP with uncertain travel costs. The total travel cost is minimized and uncertainty is expressed as intervals.

For more details about the robust vehicle routing problems and their corresponding methods of optimizations, the reader may refer to [26]. On the other hand and to the most of our knowledge, there is no study about solving the arc routing problems with the robust optimization method. In the next section, we introduce a brief review about the capacitated arc routing problem under uncertain environment and we present a mathematical formulation for the robust sparse capacitated arc routing problem under travel costs uncertainty.

3   Capacitated Arc Routing Problem Under Uncertainty

Throughout the literature, the Uncertain Capacitated Arc Routing Problem has been characterized by four stochastic factors: (1) the presence of tasks, (2) the demands of the tasks, (3) the services costs and (4) the availability of a path between each pair of vertices. Most research works consider these factors separately or combine at most two of them together, for example the presence and demands of the tasks are combined in [4]. Later, these four stochastic factors of the problem were combined all together in [20] where the authors consider the uncertainty of each of these factors with random variables and by a certain probability distribution as a function of an environmental parameter. Moreover, they introduce a mathematical formulation for the problem. The developed algorithms by them showed excellent performance for static CARP, however they were not able to find robust solutions for the uncertain CARP. Therefore, we aim in this work to design new algorithms that can find more robust solutions.

3.1 Robust Capacitated Arc Routing Problem Under Sparse Network and Under Travel Costs Uncertainty

The uncertain capacitated arc routing problem that we study is characterized by the uncertainty in the travel costs and by the sparse network for which it is defined over. This uncertainty is represented by a finite set of scenarios where each required edge of the network has a different cost with respect to each scenario. We aim at determining a robust solution i.e. in an uncertain environment, the problem objective is no longer to find a single global optimal solution, but to find a solution with the best expected quality under all possible environments. In the following, we present a mathematical modeling of the capacitated arc routing problem over sparse graph under travel costs uncertainty. In other terms, we are concerned with the Multiple-Scenario Min-Max Capacitated Arc Routing Problem under sparse graphs.

Throughout the following, let $G = (V, E)$ be a graph where V denotes the set of vertices and E the set of edges. Denote by $R \subseteq E$ the set of the required edges i.e. the set of edges having strictly positive demands to be serviced.

Lets consider the following notations and variables:

- K: the total number of the vehicles.
- Q: the capacity of each vehicle.
- $dem(e)$: the demand of the edge e.
- $\Delta_{accum}(e)$: the total demand served by the vehicle arriving at the service e including the demand of e itself which is by definition less than or equal to Q.
- c_e^s: the cost of the edge e in the scenario s.
- $N(e)$: the neighborhood of the edge e.
- $S = \{1, 2, \ldots, P\}$: the set of scenarios.
- ω_e: the capacity of edge e i.e. the maximum number of times for which an edge can be traversed.
- $x_{e,f}^{e',f'}$: a binary variable which is equal to 1 if and only if the service at f' is successive to the service at e' by the same vehicle, and the chosen shortest path between e' and f' includes the consecutive adjacent edges e and f, and 0 otherwise.
- $y_{e',f'}$: a binary variable equal to 1 if f' is serviced directly after e', and 0 otherwise.

Recall that the graphs which we are working over are sparse with maximum degree equal to 3. Denote by 0 and 1 two incident edges to the depot node. The edge 0 denotes the edge of departure of the vehicle i.e. exit from the depot, and 1 denotes the edge of returning back of the vehicle i.e. entrance to the depot after accomplishing all the services of the corresponding vehicle. Moreover, we assume that these edges are required but with a null demand. In the following, we detail the mathematical model of the problem.

The Mathematical Formulation:

$$\min_{s \in S} \max \sum_{e',f' \in R, e \in E, f \in N(e)} c_e^s x_{e,f}^{e',f'} \tag{1}$$

Subject to:

$$y_{e',f'} + y_{f',e'} \le 1 \quad \forall e', f' \in R \tag{2}$$

$$\sum_{f' \in R} y_{0,f'} = K \tag{3}$$

$$\sum_{f' \in R} y_{1,f'} = 0 \tag{4}$$

$$\sum_{e' \in R} y_{e',1} = K \tag{5}$$

$$\sum_{f' \in R} y_{e',0} = 0 \tag{6}$$

$$\sum_{f' \in R} y_{e',f'} = 1 \quad \text{if} \quad e' \neq \{0,1\} \tag{7}$$

$$\sum_{e' \in R} y_{e',f'} = 1 \quad \text{if} \quad f' \neq \{0,1\} \tag{8}$$

$$\Delta_{accum}(f') \ge \Delta_{accum}(e') + dem(f') + (dem(f') + Q) \times (y_{e',f'} - 1) \quad \forall e', f' \in R, e' \neq f' \tag{9}$$

$$\sum_{f \in N(e)} x_{e,f}^{e',f'} - \sum_{f \in N(e)} x_{f,e}^{e',f'} = 0 \quad \text{if} \quad e \neq e', e \neq f', e', f' \in R \tag{10}$$

$$\sum_{f \in N(e)} x_{e,f}^{e',f'} - \sum_{f \in N(e)} x_{f,e}^{e',f'} = y_{e',f'} \quad \text{if} \quad e = e', e', f' \in R \tag{11}$$

$$\sum_{f \in N(e)} x_{e,f}^{e',f'} - \sum_{f \in N(e)} x_{f,e}^{e',f'} = -y_{e',f'} \quad \text{if} \quad e = f', e', f' \in R \tag{12}$$

$$\sum_{e',f' \in R, f \in N(e)} x_{e,f}^{e',f'} \le \omega_e \quad \text{with} \quad \omega_e \ge 1 \quad \text{if} \quad e \neq 0 \tag{13}$$

$$\sum_{e',f' \in R, f \in N(0)} x_{0,f}^{e',f'} = K \tag{14}$$

$$\sum_{e',f' \in R, f \in N(e)} x_{f,e}^{e',f'} \le \omega_e \quad \text{with} \quad \omega_e \ge 1 \quad \text{if} \quad e \neq 1 \tag{15}$$

$$\sum_{e',f' \in R, f \in N(1)} x_{f,1}^{e',f'} = K \tag{16}$$

$$x_{e,f}^{e',f'}, y_{e',f'} \in \{0,1\}, \Delta_{accum}(e') \le Q \quad \forall e', f' \in R, e, f \in E \tag{17}$$

The objective function (1) aims to minimize the total costs under the worst case. Constraints (2) are trivial to show that either e' is serviced before f' or vice versa. Constraints (3) to (6) show that all the vehicles must depart from the depot and all the vehicles must return back to the depot after serving the required edges. The number of predecessors and the number of successors is given by the constraints (7) and (8). Constraints (9) assure that if f' is served directly after e', then the total demand done at the level of f' is greater than or equal to the total demand done at e'. Otherwise, the difference between these demands is less than Q which is trivial. Shortest path constraints are represented from (10) to (12). Constraints (13) to constraints (16) determine the capacity of each edge in G i.e. the maximum number of times an edge can be traversed, where this capacity is some ω for edges different from depot, and it is K for the edges which are incident to the depot to assure the passage of all the vehicles from and into the depot. Decision variables constraints are given in (17).

4   Efficient Algorithms for Solving Robust CARP

In this section, we present a heuristic algorithm for solving the robust sparse capacitated arc routing problem under travel costs uncertainty. The initial solution which is obtained by this algorithm is then ameliorated by a well adapted tabu search algorithm.

4.1   A Heuristic Algorithm for Solving the Robust Sparse CARP Under Travel Costs Uncertainty

This heuristic ends with a feasible initial solution of the problem. The procedure locates a worst scenario $\bar{S}$ and computes $Z(\bar{X}) = \max_x \sum c_e^{\bar{S}} x_{e,f}$. Let $e_1, e_2, \ldots, e_r$ be the required edges, and denote by λ_i the efficiency of each edge e_i which is given by the formula

$$\lambda_i = \frac{\sum_{s \in S} c_{e_i}^s}{d_{e_i}}, \tag{18}$$

where d_{e_i} denotes the demand of the required edge e_i. This algorithm is valid for the two cases of $\omega = 1$, where each edge can be traversed one only time, and $\omega > 1$, where there is a constant maximal number for traversing an edge. The only difference between the two cases lies mainly in the procedure $Update$.

The Algorithm GH.

Input: A robust CARP instance with some small density

Output: A feasible solution $\bar{X}$ and the corresponding worst scenario $\bar{S}$

Initialization

1. $|V| = n$, $E = R \cup NR$; $|E| = n + \alpha$ $(0 \leq \alpha \leq \frac{n}{2})$;
   $R = \{e_1, e_2, e_3, \ldots, e_{r-1}, e_r\}$;
   $NR = \{f_1, f_2, \ldots, f_{m-r}\}$;
2. FOR $j = 2$ to $r - 1$ DO
   Sort the required edges in the non-decreasing order of the efficiencies λ_{e_j};
   END_FOR
3. Set $\bar{X} \leftarrow 0$; $Z(\bar{X}) = +\infty$; $\bar{S} = 1$

Main Steps

1. $k \leftarrow 1$.
2. WHILE $(R <> \phi)$ and $k <= K$) DO
3. $\quad \Delta \leftarrow d_{e_2}$; $j \leftarrow 3$; $P(1) \leftarrow e_1$; $P(2) \leftarrow e_2$; $i \leftarrow 3$; $P \leftarrow \{e_1, e_2\}$;
4. $\quad$ WHILE $((\Delta <= Q)$ and $j < r)$ Do
5. $\quad\quad Subpath(e_j)$;
6. $\quad$ END_WHILE
7. $\quad P(i) \leftarrow e_r$;
8. $\quad \dim P \leftarrow i$;
9. $\quad l(i) \leftarrow \dim P$;
10. $\quad Complete(P)$;
11. $\quad \bar{X} \leftarrow \bar{X} \cup P$;
12. $\quad$ Let $\bar{S} = \max\limits_{1 \leq s \leq |S|} \{Z^s(\bar{X})\}$;
13. $\quad Update(R, NR)$;
14. $\quad k \leftarrow k + 1$;
15. END_WHILE
16. EXIT with a feasible solution $\bar{X}$ and with the corresponding worst scenario $\bar{S}$

Fig. 1. A Greedy Heuristic Algorithm GH for determining a starting feasible solution of the Robust CARP [29].

To build the solution, we used the service scheduling as to schedule which services are to be scheduled regarding each vehicle for each destination. This can be viewed in the solution representation where the services are arranged according to their order of being done as shown in the following figure, see [29].

```
1. IF (Δ + d_{e_j} <= Q) THEN
2.      Δ ← Δ + d_{e_j};
3.      P(i) ← e_j;
4.      P ← P ∪ {e_j};
5.      i ← i + 1;
6. END_IF
7. j ← j + 1;
```

Fig. 2. The Subpath procedure [29].

```
1. WHILE (l(i) > 1) DO
2.      IF (P(l) and P(l − 1) are not adjacent) THEN
3.          Dijkstra(P(l), NR, P(l − 1));
4.          P' ← Dijkstra(P(l), NR, P(l − 1));
5.          P ← P ∪ P';
6.          l(i) ← l(i) − 1;
7.      END_IF
8. END_WHILE
```

Fig. 3. The Complete procedure [29].

```
1.  FOR (i = 2 to dim P − 1) DO
2.      FOR (j = 1 to |NR|) DO
3.          IF (P(i) = (NR)_j) THEN
4.              ω((NR)_j) ← ω((NR)_j) − 1;
5.          END_IF
6.      END_FOR
7.      FOR (j = 2 to r − 1) DO
8.          IF(P(i) = e_j) THEN
9.              β ← ω(e_j);
10.             a ← e_j;
11.             FOR (l(i) = j to r − 1) DO
12.                 e_{l(i)} ← e_{l(i)+1};
13.             END_FOR
14.             |NR| ← |NR| + 1;
15.             NR(|NR|) ← a;
16.             ω(a) ← α − 1;
17.             r ← r − 1;
18.         END_IF
19.     END_FOR
20. END_FOR
```

Fig. 4. The Update procedure for $\omega > 1$ [29].

```
1.  FOR (i = 2 to dim P − 1) DO
2.      FOR (j = 1 to |NR|) DO
3.          IF (P(i) = (NR)_j) THEN
4.              For (l(i) = j to |NR|) Do
5.                  (NR)_{l(i)} ← (NR)_{l(i)+1};
6.              END_FOR
7.              |NR| ← |NR| − 1;
8.          END_FOR
9.      END_FOR
10.     FOR (j = 2 to r − 1) DO
11.         IF(P(i) = e_j) THEN
12.             FOR (l(i) = j to r − 1) DO
13.                 e_{l(i)} ← e_{l(i)+1};
14.             END_FOR
15.             r ← r − 1;
16.         END_IF
17.     END_FOR
18. END_FOR
```

Fig. 5. The Update procedure for $\omega = 1$ [29].

A solution of the problem is formed first of the services according to their order of being serviced, then we apply Dijkstra algorithm to determine a shortest path between each couple of services as shown in Fig. 6.

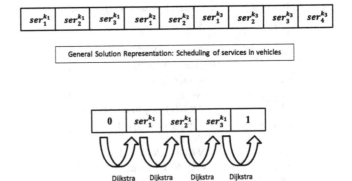

Fig. 6. An example of the solution construction.

In the example of Fig. 6, we have 3 vehicles k_1, k_2, and k_3 and each vehicle has its corresponding services. For example, for vehicle k_1, there are the services $serv_1, serv_2$, and $serv_3$ in which they are served in this order.

The vehicle k_1 departs from the depot 0 and reaches to $serv_1$ by applying Dijkstra, then it applies Dijkstra to reach to $serv_2$ and serves it, and so on until its capacity cannot be violated, then it moves from the last service to the depot again represented by 1.

In the following, we explain the main steps of the GH algorithm (Fig. 1).

- **Step 1:** k ← 1 to start with the first vehicle.
- **Step 2:** while the set of the required edges is not empty, and the number of the vehicles is less than or equal to the number of the available ones.
- **Step 3:** the total accumulated demand is fixed at the demand of the first required edge with strictly positive demand and greatest efficiency. The first edge in the path is the depot. The second edge in the path is the first required edge with a strictly positive demand.
- **Step 4:** while the accumulated demand respects the capacity of the vehicle and there are still required non-serviced edges.
- **Step 5:** call the procedure $Subpath()$, Fig. 2, that tests if adding the jth required edge will not violate the capacity constraint. In this case, update the accumulated demand to be the last one added to the demand of j, and place this edge in the ith rank of the constructed subpath. Then, move to the next rank and then to the next required edge.
- **Step 7 to Step 9:** Once adding a required edge could violate the capacity, go back to the depot. The dimension of the constructed subpath is i where the depot entrance is the ith-edge of this subpath i.e. the last edge. Assignment of the dimension of P to an auxiliary variable $l(i)$.
- **Step 10:** call the procedure $Complete$, Fig. 3. The procedure $Complete$ tests if the predecessor of each required edge in P say at rank $l(i)$ is not the required edge served in this path P and placed at the rank $l(i) - 1$, then call $Dijkstra$ and insert a shortest path of made of edges of NR between the edge at rank $l(i)$ and the edge at rank $l(i) - 1$. Each time an edge is inserted, the dimension of the path P is incremented by 1 and each edge at rank $j - 1$ will be at the rank j to let the inserted edge compensate the emptied rank.
- **Step 11:** Update the constructed solution.
- **Step 12:** Determine the worst scenario that corresponds to the scenario giving the maximal solution cost.
- **Step 13:** Update the sets NR and R: in the case where $\omega > 1$, Fig. 4, the capacity ω of any used edge in the path P is decremented by 1 each time the edge is used. The same is applied to the edges of the set R with the additional step that will be impose the removal of the used edges from R and then added to NR. For the case of $\omega = 1$, Fig. 5, the edges of both sets are removed once traversed or served.
- **Step 14:** Move to the next vehicle.

We determine by this heuristic algorithm an initial robust solution of the problem and a corresponding worst scenario.

4.2   An Adapted Tabu Search Algorithm for Solving the Robust Sparse CARP Under Travel Costs Uncertainty

In this part, we develop an adapted tabu search algorithm for the Robust Sparse CARP under travel costs uncertainty. This algorithm starts with the initial solution that is determined by the above greedy heuristic. Consider the following notations:

- X^*: best feasible solution determined by the tabu search algorithm.
- L: the tabu list.
- $Iter$: number of iteration.
- $MaxIter$: maximum number of iterations.
- $N(X^*)$: neighborhood about the solution X^*.
- S^*: the worst scenario determined by the algorithm.
- Th: a certain threshold.

The Algorithm TS.
Input: An initial feasible solution $\bar{X}$
Output: A best feasible solution X^* with the corresponding worst scenario S^*

Initialization
```
0. X* ← X̄;
1. L ← φ;
2. Iter ← 0;
```

Main Steps
```
1. WHILE (Iter < MaxIter) DO
2.    IF (|L| ≤ Th) THEN
3.       Build₁(N(X*)); /* switch 2 different services from two different vehicles */
4.    IF (|L| > Th) THEN
5.       Build₂(N(X*)); /* switch services from the same vehicle starting from some threshold */
6.    FOR (t = 1 to |N(X*)|)
7.       If ((Z(Xₜ*) ≤ Z(X*))&&(Xₜ* ∉ L)) THEN
8.          X* ← Xₜ*;
9.       END_IF
10.      Update(L), Update(Iter);
11.   END_FOR
12. END_WHILE
13. EXIT with the best solution X* and with the corresponding
    worst scenario S*
```

Fig. 7. A Tabu Search Algorithm TS for determining best solution of the robust CARP [29].

The algorithm contains several steps. Tabu search starts by an initial feasible solution obtained thanks to the greedy heuristic algorithm. All the visited solutions are feasible. The exploration of the solutions space is executed with some swaps. The elite solutions list is generated by improving the objective value where the worst scenario has already been identified. The core of the approach is to build neighborhoods and perform several local searches in order to reach a best solution. In **Step 2** of the main steps of Fig. 7, if a certain threshold Th is not attained, we diversify the search by using the procedure Build$_1$ to build the neighborhood. In this step, we choose randomly two vehicles, $vehicle_1$ and $vehicle_2$, and we select two services of each chosen vehicle i.e. $service_1^1$, $service_1^2$, $service_2^1$ and $service_2^2$. We check whether the swap of these services (the first service of the first vehicle with the first service of the second vehicle, and the second service of the first vehicle with the second service of the second vehicle) respects the capacity of the vehicles, and we swap them as explained. In this way, we explore the neighbors and we choose the best that minimizes the cost for the worst scenario. The search progresses by iteratively moving from the current solution to an improved solution. In **Step 4** of the main steps of this Figure, if the threshold Th is attained, we intensify the search by using the procedure Build$_2$ to build the neighborhood which allows the exchange of two services of the same vehicle. The tabu based strategy incorporates a tabu list in the selection mechanism that forbids the selection of the non-improving solution for a certain tabu tenure. Each visited explored solution is then settled in the tabu list L to not be visited again unless the tabu list reaches its expiration point i.e. the tabu status of a move is removed if it belongs to the list L and it exceeds $MaxIter$ iterations.

For the intensification and the diversification of the search, both are achieved via the procedures Build$_1$ and Build$_2$.

5   Computational Experiments

In this section, we introduce a set of computational experiments for which we apply each of these algorithms; the heuristic algorithm and the tabu search one. The benchmark of instances which we deal with has not been treated before. We have run all these instances with different number of scenarios and different densities of their corresponding networks in order to evaluate the performance of each of the presented algorithms. The proposed algorithms are coded in C++ and run on HP intel(R) Core(TM) i7 laptop (with 2.80 Ghz and 16 Go of Ram). These test problem instances are studied for the first time and their optimal solution values are not known. None of the previous algorithms in the literature deals with such type of instances. We work with networks having a maximum degree of 3 as we deal with a sparse network. Lets consider the following notations:

- WS_H: the worst scenario which is determined by the heuristic algorithm.
- $Cost_H$: the cost of the solution which is determined by the heuristic algorithm.
- CPU_H: the time needed by the heuristic algorithm to determine a solution.

- WS_T: the worst scenario which is determined by the tabu algorithm.
- $Cost_T$: the cost of the solution which is determined by the tabu algorithm.
- CPU_T: the time needed by the tabu algorithm to determine a solution.

The robust optimization that we apply via the developed algorithms allows us not only to get a robust solution, but also it gives us the worst scenario that may change upon the improvement of the solution as we observe in the following tables. In other terms, a worst scenario of a solution determined by the heuristic algorithm is not necessarily the same worst scenario of the solution obtained by the tabu algorithm after improvement. As a result, an improvement comes in two directions: (1) obtaining a better solution with a minimal cost and (2) improving the corresponding worst scenario.

In what follows, we show the tables of the numerical results that we have generated.

In Tables 1, 3, 5, 7, 9 and 11, we present sets of different families for the robust CARP instances. The number of vertices, the number of edges, the number of

Table 1. A first set of different families of the robust CARP problem instances - Group A_0.

Instance	$\lvert V(G)\rvert$	$\lvert E(G)\rvert$	$\lvert R\rvert$	S	K	Q
$1A_0$	10	13	5	10	2	30
$2A_0$	20	27	10	10	3	40
$3A_0$	50	70	25	10	5	60
$4A_0$	100	150	59	10	5	120
$5A_0$	231	331	121	10	10	130
$6A_0$	257	362	191	10	12	190
$7A_0$	307	439	260	10	12	225
$8A_0$	400	600	350	10	15	240

Table 2. Results of Group A_0 Instances by the heuristic algorithm and by the tabu search algorithm (TS).

Instance	WS_H	$Cost_H$	$CPU_H(s)$	WS_T	$Cost_T$	$CPU_T(s)$
$1A_0$	9	1105	0.001	9	1105	0.001
$2A_0$	10	3534	0.008	7	1943	1.6
$3A_0$	9	11476	0.01	6	5025	15
$4A_0$	2	27274	0.056	2	16037	41
$5A_0$	8	71083	0.241	8	46302	218
$6A_0$	4	120473	0.416	NI	NI	NI
$7A_0$	3	183109	0.723	3	126440	640
$8A_0$	2	255540	1.853	NI	NI	NI

NI: no improvement

Table 3. A second set of different families of the robust CARP problem instances - Group A.

Instance	$|V(G)|$	$|E(G)|$	$|R|$	S	K	Q
1A	10	13	5	10	2	15
2A	20	27	10	10	2	40
3A	50	70	25	10	3	50
4A	100	150	59	10	3	120
5A	231	331	121	10	4	120
6A	257	362	191	10	5	150
7A	307	439	260	10	7	150
8A	400	600	350	10	7	150

required edges (services), the number of scenarios, the number of the available homogenous vehicles and their capacity are all given. The results in Tables 2, 4, 6, 8, 10 and 12 present the worst scenario determined by the heuristic, the cost of the solution and the CPU consuming time of the heuristic too. Furthermore, it gives the worst scenario determined by the tabu search algorithm, the cost of the corresponding solution and the CPU consuming time of the tabu search algorithm.

On one hand, we notice that our greedy heuristic succeeds to have the access to all the studied instances with a very small CPU consuming time regardless the quality of the found solution. On the other hand, the tabu algorithm did not succeed to have the access to the big size instances.

Recall that G denotes a graph where $V(G)$ denotes the set of vertices and $|V(G)|$ its cardinality, $E(G)$ denotes the set of edges and $|E(G)|$ its cardinality. The set of required edges is represented by R, and the set of different scenarios is represented by S. The used fleet of vehicles is homogeneous where K denotes

Table 4. Results of Group A Instances by the heuristic algorithm and by the tabu search algorithm (TS).

Instance	WS_H	$Cost_H$	$CPU_H(s)$	WS_T	$Cost_T$	$CPU_T(s)$
1A	10	1295	0.001	10	1059	0.5
2A	1	3104	0.008	1	2644	1.52
3A	6	10169	0.012	6	5658	8.326
4A	2	27370	0.056	10	17735	40.96
5A	3	46461	0.236	3	31074	124
6A	1	84875	0.392	5	50972	208
7A	5	124241	0.664	6	76389	357
8A	5	118039	1.168	5	71498	450

Table 5. A third set of different families of the robust CARP problem instances - Group B_0.

Instance	$|V(G)|$	$|E(G)|$	$|R|$	S	K	Q
$1B_0$	10	13	5	40	2	30
$2B_0$	20	27	10	40	3	40
$3B_0$	50	70	25	40	5	60
$4B_0$	100	150	59	40	5	120
$5B_0$	231	331	121	40	10	130
$6B_0$	257	362	191	40	12	190
$7B_0$	307	439	260	40	12	225
$8B_0$	400	600	350	40	15	240

Table 6. Results of Group B_0 Instances by the heuristic algorithm and by the tabu search algorithm (TS).

Instance	WS_H	$Cost_H$	CPU_H	WS_T	$Cost_T$	CPU_T
$1B_0$	10	1273	0.001	10	1273	0.001
$2B_0$	6	2650	0.0016	38	1927	6
$3B_0$	30	9807	0.052	24	8206	33
$4B_0$	30	28178	0.204	30	15802	287
$5B_0$	8	66736	0.936	NI	NI	NI
$6B_0$	10	118841	1.668	NI	NI	NI
$7B_0$	23	178188	2.668	NI	NI	NI
$8B_0$	12	242994	4.752	NI	NI	NI

NI: no improvement

the number of the available vehicles and Q represents the capacity of each one. The different instances are generated randomly i.e. the sparse graph is generated randomly but respecting that $|E(G)| = |V(G)| + \alpha$ with $1 \leq \alpha \leq \frac{|V(G)|}{2}$ and that the maximum degree in this network is 3. The costs over the scenarios are all generated randomly too.

The numerical instances are divided into 3 groups and each group into two parts: Groups A_0 and A with 10 scenarios (Tables 1 and 3), Groups B_0 and B with 40 scenarios (Tables 5, 7) and Groups C_0 and C with 100 scenarios (Tables 9 and 11). The reason beyond this decomposition of instances comes after we have noticed that not all the available vehicles are used in the solution, thus we generated instances with a smaller number of vehicles to observe whether this factor may affect the solution.

Table 7. A fourth set of different families of the robust CARP problem instances - Group B.

Instance	$|V(G)|$	$|E(G)|$	$|R|$	S	K	Q
1B	10	13	5	40	2	15
2B	20	27	10	40	2	40
3B	50	70	25	40	3	50
4B	100	150	59	40	3	120
5B	231	331	121	40	4	120
6B	257	362	191	40	5	150
7B	307	439	260	40	7	150
8B	400	600	350	40	7	150

Table 8. Results of Group B Instances by the heuristic algorithm and by the tabu search algorithm (TS).

Instance	WS_H	$Cost_H$	CPU_H	WS_T	$Cost_T$	CPU_T
1B	20	1607	0.002	18	1529	2.216
2B	25	3594	0.052	15	1969	5.944
3B	1	9989	0.06	33	5439	35.152
4B	39	28120	0.305	37	16389	169
5B	22	43603	0.852	NI	NI	NI
6B	21	87199	1.723	NI	NI	NI
7B	8	119613	2.683	NI	NI	NI
8B	31	152818	2.8	NI	NI	NI

NI: no improvement

Table 9. A fifth set of different families of the robust CARP problem instances - Group C_0.

Instance	$|V(G)|$	$|E(G)|$	$|R|$	S	K	Q
$1C_0$	10	13	5	100	2	30
$2C_0$	20	27	10	100	3	40
$3C_0$	50	70	25	100	5	60
$4C_0$	100	150	59	100	5	120
$5C_0$	231	331	121	100	10	130
$6C_0$	257	362	191	100	12	190
$7C_0$	307	439	260	100	12	225
$8C_0$	400	600	350	100	15	240

Comments: Consider the instances of Groups A_0 and A where we have a relatively small number of scenarios (10 scenarios). It is obvious that both algorithms perform well whatever the size of the instance is i.e. Instances $1A_0$, $1A$, $2A_0$, $2A$, $3A_0$ and $3A$ which are considered as small size instances are solved rapidly (Tables 2 and 4). The gap between the solution determined by the heuristic and that determined by the tabu is relatively high (between 40% and 80%), and this shows that the corresponding tabu search algorithm is efficient and it can ameliorate the solution very well. We have to recall here that we aim at determining a robust solution for the problem despite of the high quality of this solution. Medium size instances (Instances $4A_0$ and $4A$, Tables 2 and 4) need a small CPU consuming time to be solved by the heuristic, while they require more time to be solved by the tabu. Big size instances (Instances $5A_0$, $5A$, $6A_0$, $6A$, $7A_0$, $7A$, $8A_0$ and $8A$, Tables 2 and 4) are solved rapidly by the heuristic just like the other instances, while it is not the case for the tabu search algorithm that needs more time to perform. As a brief conclusion for the first group of instances, the greedy

Table 10. Results of Group C_0 Instances by the heuristic algorithm and by the tabu search algorithm (TS).

Instance	WS_H	$Cost_H$	CPU_H	WS_T	$Cost_T$	CPU_T
$1C_0$	62	1250	0.016	62	1250	0.016
$2C_0$	89	2746	0.028	89	1767	28
$3C_0$	73	12762	0.144	29	6389	153
$4C_0$	81	27257	0.472	29	15634	731
$5C_0$	6	72386	2.264	NI	NI	NI
$6C_0$	2	112454	3.94	NI	NI	NI
$7C_0$	95	189989	6.556	NI	NI	NI
$8C_0$	10	296212	12.196	NI	NI	NI

NI: no improvement

Table 11. A sixth set of different families of the robust CARP problem instances - Group C.

Instance	$|V(G)|$	$|E(G)|$	$|R|$	S	K	Q
Instance 1C	10	13	5	100	2	15
Instance 2C	20	27	10	100	2	40
Instance 3C	50	70	25	100	5	60
Instance 4C	100	150	59	100	5	120
Instance 5C	231	331	121	100	7	120
Instance 6C	257	362	191	100	12	190
Instance 7C	307	439	260	100	12	225
Instance 8C	400	600	350	100	15	240

Table 12. Results of Group C Instances by the heuristic algorithm and by the tabu search algorithm (TS).

Instance	WS_H	$Cost_H$	CPU_H	WS_T	$Cost_T$	CPU_T
1C	48	1548	0.007	48	1216	4.872
2C	49	3655	0.27	49	2520	15.556
3C	32	9924	0.241	79	5689	101.084
4C	42	27870	0.452	54	18180	441.576
5C	66	45412	1.624	NI	NI	NI
6C	44	91616	3.81	NI	NI	NI
7C	13	149114	5.318	NI	NI	NI
8C	79	108750	8.375	NI	NI	NI

NI: no improvement

heuristic algorithms behaves almost in the same way for all the instances where it determines a solution of the problem within a very small consuming time. However, as the size of the studied problem instance increases, the consuming time needed by the tabu search algorithm increases too, though it ameliorates the quality of the solution obviously. Concerning the effect of the number of the available vehicles K on the solution, we observe that for some big size instances, as K decreases, it becomes easier to ameliorate the initial solution as seen for instances $6A_0, 8A_0, 6A$ and $8A$. Moreover, the CPU consuming time of the tabu algorithm decreases with the decrease of K.

For a medium number of scenarios (40 scenarios) represented by Groups B_0 and B instances (Tables 5 and 7), we observe that the performance of the heuristic algorithm is almost the same for all the instances (Tables 6 and 8). The performance of the tabu search algorithm differs according to the size of the instance i.e. as the number of the vertices of the network increases, the CPU consuming time of this algorithm increases too. Though there is a high improvement of the quality of the solution. However, we see that there is no rapid improvement for big size instances (Tables 6 and 8).

Concerning the last group of instances; Groups C_0 and C with a big number of scenarios (100 scenarios, Tables 9 and 11), the heuristic algorithm performs rapidly and determines a solution within a very short time for all the instances, while the performance of the tabu search algorithm is affected by the size of the instance and it needs more time and memory to improve the solution found by the heuristic (Tables 10 and 12). Furthermore, as the number of available vehicles decreases, the tabu algorithm performs faster for the small size instances, whereas it fails to improve for the medium and big size instances.

A general conclusion is drawn out. On one hand, the heuristic algorithm is able to determine an initial solution for any problem instance and for any number of scenarios within a very short CPU consuming time. On the other hand, the performance of the developed tabu search algorithm is related to the number of scenarios and to the size of the studied problem instance. In other

172 S. Tfaili et al.

terms, as the number of scenarios increases and as the size of the studied problem instance increases, the CPU consuming time of the tabu search algorithm increases too. However, this algorithm is able to improve very well the solution which is determined by the heuristic whenever it is able to improve.

6 Conclusion

Recent studies in the domain of routing problems are subjected towards the robust optimization of these problems such as the robust vehicle routing problem. In this paper we study the robust sparse capacitated arc routing problem under travel costs uncertainty. As real life applications may face a high degree of uncertainty which stands against the feasibility or the optimality of the solution, it is important to find a robust solution whatever the situation is. We choose the robust optimization to address the travel costs uncertainty and we represent this uncertainty by a set of scenarios. We solve the problem by a min-max approach that it by minimizing the solution under the worst scenario. A mathematical modeling of the robust problem is given and two algorithms are developed. A greedy heuristic algorithm is constructed first to determine an initial feasible solution of the problem, and then an adapted tabu search algorithm is developed. The tabu search algorithm starts with the initial solution determined by the greedy heuristic and tries to improve the quality of this solution.

One more time we generate our own benchmark of instances respecting the structure of the graphs which we are working over. The computational experiments show the high effectiveness and robustness of the heuristic which is able to find a feasible initial solution whatever the size of the instance is in a very small CPU consuming time. The tabu algorithm improves the quality of the solution obtained by the heuristic significantly for the small and medium size instances, whereas it does not succeed to ameliorate that of the big size instances. However, the main objective is to determine a robust solution with best expected quality under several possible conditions and not to find a single global optimal solution.

As a future work, we aim at studying the robust sparse capacitated arc routing problem under other forms of uncertainty such as demands uncertainty as well at developing other algorithms to solve the robust problem such as genetic algorithm which is worth to be tested. Another potential future research work would be to use the high performance computing as known as HPC to solve very large size instances with many scenarios and very small density. We think that our approach if implemented with HPC would able to obtain very competitive solutions.

References

1. Álvarez-Miranda, E., Ljubić, I., Toth, P.: Exact approaches for solving robust prize-collecting Steiner tree problems. Eur. J. Oper. Res. **229**(3), 599–612 (2013)
2. Ben-Tal, A., Ghaoui, L.E., Nemirovski, A.: Robust Optimization. Princeton University Press, Princeton (2009)

3. Bertsimas, D.: Probabilistic combinatorial optimization problems. Ph.D. thesis Massachusetts Institute of Technology, Dept. of Mathematics (1988)
4. Bertsimas, D.: A vehicle routing problem with stochastic demand. Oper. Res. **40**(3), 574–585 (1992)
5. Bertsimas, D., Gamarnik, D., Rikun, A.A.: Performance analysis of queueing networks via robust optimization. Oper. Res. **59**(2), 455–466 (2011)
6. Bertsimas, D., Sim, M.: Robust discrete optimization and network flows. Math. Program. **98**, 49–71 (2003)
7. National Center for Biotechnology Information. http://www.ncbi.nlm.nih.gov
8. Christiansen, C.H., Lysgaard, J.: A branch-and-price algorithm for the capacitated vehicle routing problem with stochastic demands. Oper. Res. Lett. **35**(6), 773–781 (2007)
9. Christiansen, C.H., Lysgaard, J., Wøhlk, S.: A branch-and-price algorithm for the capacitated arc routing problem with stochastic demands. Oper. Res. Lett. **37**(6), 392–398 (2009)
10. Coco, A.A., Solano-Charris, E.L., Santos, A.C., Prins, C., Noronha, T.: Robust optimization criteria: state-of-the-art and new issuses. Technical report UTT-LOSI -14001. ISSN:2266–5064. Université de Technologie de Troyes (2014)
11. Dantzig, G.B., Ramser, J.H.: The truck dispatching problem. Manage. Sci. **6**(1), 80–91 (1959)
12. Fleury, G., Lacomme, P., Prins, C.: Evolutionary algorithms for stochastic arc routing problems. In: Raidl, G.R., et al. (eds.) EvoWorkshops 2004. LNCS, vol. 3005, pp. 501–512. Springer, Heidelberg (2004). https://doi.org/10.1007/978-3-540-24653-4_51
13. Han, J., Lee, C., Park, S.: A robust scenario approach for the vehicle routing problem with uncertain travel times. Transp. Sci. **48**(3), 373–390 (2013)
14. Kasperski, A., Zieliński, P.: On the approximability of robust spanning tree problems. Theoret. Comput. Sci. **412**(4–5), 365–374 (2011)
15. Kenyon, A.S., Morton, D.: Stochastic vehicle routing with random travel times. Transp. Sci. **37**(1), 69–82 (2003)
16. Kouvelis, P., Yu, G.: Robust Discrete Optimization and its Applications. Kluwer Academic Publishers, Norwell (1997)
17. Laporte, G., Louveaux, F., Mercure, H.: The vehicle routing problem with stochastic travel times. Transp. Sci. **26**(3), 161–170 (1992)
18. Laporte, G., Louveaux, F., Hamme, L.V.: An integer L-shaped algorithm for the capacitated vehicle routing problem with stochastic demands. Oper. Res. **50**(3), 415–423 (2002)
19. Mattia, S.: The robust network loading problem with dynamic routing. Comput. Optim. Appl. **54**(3), 619–643 (2013)
20. Mei, Y., Tang, K., Yao, X.: Capacitated arc routing problem in uncertain environments. In: IEEE World Congress on Computational Intelligence, pp. 1400–1407 (2010)
21. Monaci, M., Pferschy, U., Serafini, P.: Exact solution of the robust knapsack problem. Comput. Oper. Res. **40**(11), 2625–2631 (2013)
22. Sbihi, A.: A cooperative local serach-based algorithm for the multiple-scenario max-min knapsack problem. Eur. J. Oper. Res. **202**(2), 339–346 (2010)
23. Secomandi, N.: Comparing neuro-dynamic programming algorithms for the vehicle routing problem with stochastic demands. Comput. Oper. Res. **27**(11–12), 1201–1225 (2000)
24. Secomandi, N., Margot, F.: Reoptimization approaches for the vehicle-routing problem with stochastic demands. Oper. Res. **57**(1), 214–230 (2009)

25. Shapiro, A., Dentcheva, S., Ruszczyński, A.: Lectures on Stochastic Programming: Modeling and Theory, vol. 9 (2008)
26. Solano Charris, E.L.: Optimization methods for the robust vehicle routing problem. In: Ph.D. thesis, Université de Technologie de Troyes (2015)
27. Sun, L.: A new robust optimization model for the vehicle routing problem with stochastic demands. J. Interdisc. Math. **17**(3), 287–309 (2014)
28. Sungur, I., Ordóñez, F., Dessouky, M.: A robust optimization approach for the capacitated vehicle routing problem with demand uncertainty. IIE Trans. **40**(5), 509–523 (2008)
29. Tfaili, S., Sbihi, A., Yassine, A., Diarrassouba, I.: Capacitated arc routing problem over sparse underlying graph and under travel costs uncertainty. In: Proceedings of the 7th International Conference on Operations Research and Enterprise Systems: ICORES, vol. 1, pp. 144–151 (2018)
30. Toklu, N.E., Montemanni, R., Gambardella, L.M.: An ant colony system for the capacitated vehicle routing problem with uncertain travel costs. In: IEEE Symposium on Swarm Intelligence (SIS), pp. 32–39 (2013)
31. Toklu, N.E., Montemanni, R., Gambardella, L.M.: A robust multiple ant colony system for the capacitated vehicle routing problem. In: IEEE International Conference on Systems, Man, and Cybernetics (SMC), pp. 1871–1876 (2013)
32. Verweij, B., Shabbir, A., Kleywegt, A.J., Nemhauser, G., Shapiro, A.: The sample average approximation method applied to stochastic routing problems: a computational study. Comput. Optim. Appl. **24**(2–3), 289–333 (2003)
33. Waters, C.: Vehicle-scheduling problems with uncertainty and omitted customer. J. Oper. Res. Soc. **4**(12), 1099–1108 (1989)
34. Wets, R.J.B.: Stochastic programming models: wait-and-see versus here-and-now. In: Greengard, C., Ruszczynski, A. (eds.) Decision Making Under Uncertainty. The IMA Volumes in Mathematics and its Applications, vol. 128, pp. 1–15. Springer, New York (2002). https://doi.org/10.1007/978-1-4684-9256-9_1
35. Mei, Y., Tang, K., Yao, X.: Capacitated arc routing problem in uncertain environments. In: Proceedings of the 2010 IEEE Congress on Evolutionary Computation, pp. 1400–1407 (2010)

Applications

Optimal Air-Cargo Allotment Contract with Multiple Freight Forwarders

Kannapha Amaruchkul$^{(\boxtimes)}$

Graduate School of Applied Statistics,
National Institute of Development Administration (NIDA),
118 Serithai Road, Bangkok 10240, Thailand
kamaruchkul@gmail.com

Abstract. Consider the air-cargo service chain which comprises a carrier and multiple forwarders. The carrier and each of the forwarders may establish an allotment contract at the start of the season. We formulate the contract design problem as a Stackelberg game, in which the carrier is the leader and offers a contract to a forwarder. The contract parameters may include the discount contract price and the penalty cost for the unused allotment as well as the minimum allotment utilization. The carrier's contract is accepted, if the forwarder earns at least its reservation profit. Given the carrier's offer, the forwarder decides how much to book as an allotment, in order to maximize its own expected profit. We show that the two-parameter contract suffices to coordinate the service chain, and the carrier earns the maximum chain's expected profit less the total reservation profits of all forwarders. If the penalty cost is not imposed, then the minimum allotment utilization is needed to construct an efficient contract. On the other hand, if the penalty cost is strictly positive, then there is no need to impose the minimum allotment requirement.

Keywords: Air-cargo · Capacity management · Stochastic model applications

1 Introduction

Air-cargo operations inarguably play a crucial role in the modern supply chain, since they improve efficiency in logistics and increase competitive advantages. Despite the 1% world trade by volume, airfreight represents more than 35% of global trade by value [23]. The air-cargo growth is driven by global liberalization, cross-border e-commerce, and the implementation of supply chain/logistics management strategies, which emphasize on short lead times, e.g., lean/agile management and just-in-time (JIT) production systems. With e-commerce boom, airfreight has become a *de facto* mode of cross-border transportation, for the customer centric businesses with fast delivery times. Air cargo consists of various commodity types, e.g., pharmaceutical products, live animals, electronic devices, human remains, dry ice, and gold bullion; some fastest-growing air-cargo perishables in 2017 include seafood from Scotland, smoked meat and wines from

© Springer Nature Switzerland AG 2019
G. H. Parlier et al. (Eds.): ICORES 2018, CCIS 966, pp. 177–197, 2019.
https://doi.org/10.1007/978-3-030-16035-7_10

Australia, clotted cream from the U.K., blueberries from Ukraine and medicinal plants from Afghanistan [40]. Despite the sluggish growth in 2015 due to the economy slowdown worldwide, air-cargo traffic is expected to gradually accelerate. The largest average annual growth rate is found in Asia-Pacific freight market [11].

In an air-cargo service chain, a shipper can receive services directly from an air-cargo carrier or delegate to a freight forwarder. A large portion of air cargo volume is handled through freight forwarders. A freight forwarder acts as an intermediary party, who connects a shipper to an airline. The forwarder consolidates shipments and handles various aspects of the shipping process, e.g., pickup and delivery services, insurance, customs clearance, import and export documentation, cargo tracking and tracing, and interacting with multi-modal carriers. Most forwarders do not own an airplane and obtain cargo space on *ad hoc* basis or through a medium- or long-term capacity agreement, also known as the *allotment*, with the carrier. The airline carries consolidated cargo in the belly of a passenger plane or a freighter. Freighters are critical to compete in air cargo markets, since they carry more than half of air-cargo traffic and airlines operating freighters generate 90% of the industry revenues [11]. Capacity utilization is one of the top operational problems, faced by the majority of the cargo carriers [1]. The carrier offers a contract to the forwarder, hoping to increase its capacity utilization. The forwarder wants to establish the contract, in order to receive volume discounts or lower freight charges. The discount may depend on the size of the allotment and the actual volume tendered by the forwarder [34].

Air-cargo spaces are perishable in the sense that they cannot be sold after the flight departure. They are sold in two stages: In the first stage which happens a few months before a season starts, a carrier allocates spaces to forwarders either as part of a binding contract or as part of goodwill [10]. Each year comprises two seasons, Winter and Summer schedules, specified by the International Air Transport Association [35]. Through the allotment contract, the forwarder achieves a more economical rate, compared to the so-called *spot rates* for *ad hoc* shipments. The forwarder pre-books a certain amount of capacity at a pre-specified rate, based on its anticipated demand on a given route and the contract terms. The demand is forecasted based various factors such as the economic condition, the competitors' action, and the projected trend and seasonal patterns. About 50–70% of air-cargo space is sold to forwarders through a "hard" block space agreement (BSA) at a negotiated price, a "soft" block permanent booking (PB) or other forms of capacity agreements [33]. Carriers in Asia Pacific typically allocate a large fraction, whereas those in North America allocate a small fraction of their capacity [19]. After the forwarder collects and consolidates shipments from its customers, and the actual allotment usage becomes known, the payment is transferred between the carrier and the forwarder. If the forwarder's customer demand is smaller than previously anticipated, the allotment utilization by the forwarder may be low. The carrier may impose some cancellation fee for the unused allotment by the forwarder, or it may impose the minimum allotment utilization and offer the refund up to a pre-specified portion, not all unused portion of the allotment. Nevertheless, for the airline's most important forwarders, the cancellation clause is rarely enforced; these powerful forwarders pay only for their actual allotment usages. After the unused allotment is released by the forwarder a few weeks before a flight departure, the carrier re-sells the remaining capacity on a *free-sale* or *ad hoc* basis to direct shippers.

In this article, we develop a formulation for the study of contracts with three parameters: (1) a discount contract price, (2) a refund (or penalty cost) for the unused portion of the allotment, and (3) the minimum allotment utilization. Our objective is to determine an optimal contract scheme, which allows the air-cargo service chain to be efficient. To this end, we formulate a Stackelberg game, in which the carrier is the leader and proposes a contract to each of the multiple freight forwarders. Based on its anticipated demand and the contract parameters, the forwarder determines the best allotment size, which maximizes its own expected profit. Based on the forwarder's best response, the carrier determines the contract parameters in order to maximize its own expected profit. We analyze the sequential game of the allotment contract problem and identify sufficient conditions under which the equilibrium contract is efficient. Our model benefits the carrier by identifying a possible contract structure that it should strive for in negotiating with the forwarders. The contract with only two parameters (either the positive penalty cost or the minimum allotment utilization) is sufficient to coordinate the air-cargo service chain.

Air-cargo capacity is perishable and can be sold at different prices to heterogeneous customers with different willingness to pay. Thus, it is a prime candidate for applying revenue management (RM) strategies. Overview of RM theory and practice can be found in textbooks, e.g., [21,31,36,43], and journal articles, e.g., [13,27,30]. Literature on air-cargo RM is fairly limited, in comparison to the extensive literature on passenger RM. [16] provides a literature review on air-cargo operations. [10,24] are among the early descriptive overview papers on air-cargo RM. [8] describes the air-cargo system in the Asia Pacific. [35] describes the implementation of air-cargo RM system at KLM and highlights key factors that critically affect its performance. Air-cargo operations are presented in [33], and air-cargo RM from business perspective is discussed in [9,14]. The air-cargo industry outlook can be found in, e.g., [11,23]. Air-cargo training courses are provided by several professional associations such as International Air Transport Association (IATA), British International Freight Association (BIFA) and International Association of Airport Executives Canada (IAAEC).

Key short-term air-cargo operations include aircraft loading (e.g., [37,41]), shipment routing (e.g., [32,42]) and booking control (e.g., [3,7,20,45]). Allotment contracts are medium-term decisions. Articles which combine both short-term booking control and medium-term allotment decisions are, e.g., [25,28,44]. [25] considers an airline which operates parallel flights between a given origin and destination pair; the carrier's medium-term decision is to choose allotment contracts among available bids from forwarders, whereas the short-term decision corresponds to the booking control problem, from which the expected contribution from the spot market can be determined. In [28], the carrier's medium-term decision is to determine how much of the total weight and volume capacity to sell as allotments. Unlike ours, these articles take the carrier's perspective and are concerned with a single decision maker.

In contrast to having a carrier as the single decision maker, the contract design problem considers two or more decision makers, which typically include the carrier and one or more forwarders, and the game theoretic approach is often employed to find an optimal allotment contract scheme. [18] proposes an options contract, similar to supply chain contracts in, e.g., electricity generation and semiconductor manufacturing, and investigates the suitability of options contracts in the air-cargo industry. Under certain contract parameters and a suitable spot market environment, the options contract outperforms the fixed-commitment contract. The buy-back scheme is another prevalent contract in supply chain (see, e.g., [12] for a review of supply contracts): [26] applies this buy-back concept in the air-cargo service chain and shows that the buy-back policy improves revenues of both players, namely the carrier and the forwarder. [4] considers the carrier's mechanism design problem, in which the other player, namely the freight forwarder, possesses some private information on, e.g., its customer demand and operating cost. An optimal allotment scheme, which maximizes the total contribution of the air-cargo service chain, is attainable via a contract with an appropriate upfront and cancellation fees. [17] provides conditions such that flexible contract schemes maximize the total profit in the service chain. [2] proposes an allotment contract, which includes a discount contract price, a penalty cost for an unused portion of the allotment, and an allotment utilization requirement, and derives a sufficient condition for an optimal contract.

This article extends [2] to include multiple freight forwarders. The carrier's capacity allocation problem with multiple forwarders is studied in [5]. The expected contribution given a fixed allotment is obtained using a discrete Markov chain, and the maximization of the total expected is formulated as a Markov decision process. [5] derives the optimal allotment from the carrier's viewpoint, given that the contractual agreement is exogenously given. [15] proposes the tying capacity allocation mechanism, in which multiple routes with different capacity utilization are included in the contract. The carrier allocates capacities to multiple forwarders using their performances on different routes in the previous year. [38] considers a multiple-forwarder setting and proposes an options contract to mitigate the carrier's capacity utilization risk. [15] solves for an optimal solution using a dynamic programming, and [38] provides a numerical example to show how to obtain an optimal contract, whereas we analytically derive a sufficient condition for an optimal contract. With the exception of [6,39], which consider risk-adverse party, these articles including ours assume that each player is rational, risk-neutral and maximizes its expected profit. These papers employ a mechanism design approach to find an optimal contract. Ours contributes to this literature: We consider a different scheme and provide a sufficient condition such that a two-parameter contract coordinates the chain.

The rest of this article is organized as follows. Section 2 presents the Stackelberg game of the interaction between the carrier and multiple freight forwarders. A sufficient condition for an equilibrium of the game is derived in Sect. 3. We also provide an analysis for the centralized chain, in which all decision makers are assumed to be owned by one single company. Section 4 provides some numerical examples to illustrate our approach, and Sect. 5 gives a summary and a few extensions.

2 Formulation

Consider an air-cargo service chain, which consists of an air-cargo carrier endowed with cargo capacity of κ and m freight forwarders, referred to as $1, 2, \ldots, m$. At the beginning of the season, each forwarder and the carrier independently negotiate the allotment contract. The freight forwarder wants to pre-book capacity in bulk with the carrier to achieve the discount rate, which is less than or equal to the spot rate. Let v_i denote the spot rate that forwarder i obtains on *ad hoc* shipments (without the allotment contract). Assume that the forwarders are labeled such $v_1 \geq v_2 \geq v_3 \cdots \geq v_m$. In the strategic level, the air-cargo capacity is assumed to be a one-dimensional quantity. In the operational level, the carrier charges the forwarder based on the chargeable weight, which is the maximum between the volume weight and the gross weight. If the shipment is measured in centimeter, then the volume weight is equal to the shipment's cubic centimeter divided by 6000. We concern with the carrier's strategic decisions, not operational.

The interaction between the carrier and the forwarders is modeled as a Stackelberg game. The sequence of events is as follows: The carrier offers forwarder i, the three-parameter contract $\Omega_i = (w_i, h_i, u_{ri})$ where $0 < w_i \leq v_i$ is the discount rate, h_i the penalty cost for unused portion of the allotment and u_{ri} is the required expected allotment utilization rate. Forwarder i either rejects or accepts the contract proposal. If the contract Ω_i is accepted, then forwarder i determines the size of the allotment, denoted by x_i. The allotment decision takes place before demands materialize. For each $i = 1, 2, \ldots, m$, let D_i denote the stochastic demand to forwarder i. During the season, forwarder i accepts all demand D_i at the per-unit price p_i where we assume that $p_i \geq v_i$. Given the allotment x_i, the expected contribution of forwarder i is as follows:

$$\pi_i(x_i, \Omega_i) = E[p_i D_i - t_i(x_i, D_i) - v_i(D_i - x_i)^+] \tag{1}$$

where the transfer payment from forwarder i to the carrier is

$$t(x_i, D_i) = w_i \min(D_i, x_i) + h_i(x_i - D_i)^+. \tag{2}$$

In (1), the first term is revenue forwarder i earns from its customer demand D_i, the second term the contract payment, and the third term the forwarder's spot purchase for the excess demand. In (2), the contract payment is the sum of the payment for the actual allotment usage and the penalty cost associated with the unused allotment.

The first two contract parameters (w_i, h_i) can be interpreted differently as follows: The transfer payment from forwarder i to the carrier can be written as

$$t_i(x_i, D_i) = w_i \min(D_i, x_i) + h_i(x_i - D_i)^+$$

$$= w_i x_i + (h_i - w_i)(x_i - D_i)^+ \tag{3}$$

$$= w_i x_i - (w_i - h_i)(x_i - D_i)^+ \tag{4}$$

In (3) and (4), we can interpret w_i as the wholesale price for the entire allotment x_i, paid upfront by the forwarder. If $h_i > w_i$, then the forwarder pays for the allotment x_i upfront at the wholesale price of w_i, and after its demand is realized, the penalty rate of $(h_i - w_i)$ is charged for the unused portion of the allotment; see (3). If $h_i < w_i$, then the forwarder pays for the allotment x_i upfront at the wholesale price of w_i as before, but after its demand is realized, the refund rate of $(w_i - h_i)$ for the unused portion of the allotment is returned from the carrier to the forwarder; see (4). In particular, the contract parameter $h_i = 0$ corresponds to the full refund; the forwarder pays for the allotment x_i upfront at the wholesale price w_i and it gets a full refund rate of w_i for the unused portion of the allotment. Finally, if $h_i = w_i$, then the forwarder pays for the entire allotment x_i at the wholesale price w_i upfront, and there are no additional monetary transfers. Since the air-cargo selling season is so short that monetary discount can be ignored, the expected profit is not affected by the timing in which the payment is collected. Our formulation subsumes both refund $(h_i < w_i)$ and penalty $(w_i < h_i)$ rates for the unused portion of the allotment. Furthermore, the full-refund contract $(h_i = 0)$ is not uncommon, especially when the freight forwarder is very powerful and holds a large market share on a route.

The third contract parameter u_{ri} ensures that forwarder i maintains the allotment utilization of at least u_{ri}; specifically,

$$u_i(x) \geq u_{ri}$$

where the allotment utilization is defined as the ratio of the expected actual allotment usage to the allotment size x:

$$u_i(x) = \frac{1}{x} E[\min(D_i, x)]. \tag{5}$$

In practice, the forwarder generally needs to maintain the high utilization; otherwise, the carrier may choose not to continue with this forwarder in the future or may not offer a favorable contract term to the forwarder.

Forwarder i accepts the contract if the expected contribution exceeds the forwarder's reservation profit, denoted by ϵ_i. Let $x_i^*(\Omega_i)$ denote forwarder i's best response to the contract parameter Ω_i; i.e.,

$$x_i^*(\Omega_i) = \operatorname{argmax}\{\pi_i(x_i, \Omega_i) : u_i(x_i) \geq u_{ri}\}. \tag{6}$$

The contract Ω_i is accepted if $\pi_i(x_i^*(\Omega_i), \Omega_i) \geq \epsilon_i$. If the reservation profit were thought of as the expected contribution if the forwarder did not have an allotment contract, then we could set $\epsilon_i = E[(p_i - v_i)D_i]$, the contribution margin in the spot market multiplied by the expected demand. This quantity can be viewed as the lower bound on the forwarder's reservation profit.

Let $\mathbf{H} = ((w_i, h_i, u_{ri}) : i = 1, 2, \ldots, m)$ denote the contract parameters offered by the carrier to forwarders $1, 2, \ldots, m$, respectively, and $\mathbf{x} = (x_i : i = 1, 2, \ldots, m)$ the allotments chosen by the forwarders. After all forwarders release their unused allotments, the carrier re-sells this to direct shippers. Let p_0 be the carrier's price charged to the direct-ship demand, denoted by D_0. The carrier's expected profit is defined as:

$$\psi(\mathbf{x}, \mathbf{H}) = E[p_0 \min(D_0, \kappa - \sum_{i=1}^{m} \min(D_i, x_i))$$

$$+ \sum_{i=1}^{m} t_i(x_i, D_i)]. \tag{7}$$

In (7), the first term is the carrier's revenue from selling the remaining cargo space to its own direct-ship customers, and the second term the sum of all forwarders' transfer payments.

At the equilibrium, the carrier anticipates the forwarders' best responses $\mathbf{x}^*(\mathbf{H}) = (x_i^*(\Omega_i); i = 1, 2, \ldots, m)$ and chooses the best contract parameters, in order to its own expected profit:

$$\max_{\mathbf{H}} \quad \psi(\mathbf{x}^*(\mathbf{H}), \mathbf{H}) \tag{8}$$

$$\text{subject to:} \quad \sum_{i=1}^{m} x_i^*(\Omega_i) \leq \kappa. \tag{9}$$

In Sect. 3, we will determine the equilibrium of the game, in which each party maximizes its own expected profit. We refer to this as the *decentralized chain*.

Finally, we consider the entire air-cargo service chain: Suppose that all forwarders and the carrier are owned by the same firm, called the *integrator*. This is referred to as the *centralized chain*. The total chain's expected profit is defined as the sum of the forwarders' expected profits and that of the carrier:

$$\tau(\mathbf{x}) = E[\sum_{i=1}^{m} \pi_i(x_i, \Omega_i) + \psi(\mathbf{x}, \mathbf{H})]$$

$$= E[\sum_{i=}^{m} (p_i D_i - v_i(D_i - x_i)^+)$$

$$+ p_0 \min(D_0, \kappa - \sum_{i=1}^{m} \min(D_i, x_i))]. \tag{10}$$

In (10), the first term is the total contribution from customers in need of dedicated services as offered by the forwarder, and the second term the contribution from the direct-ship customer. Note that in the integrator's profit, there are no contract payment terms, because we assume that the forwarders and the carrier belong to the same firm, and their transfer payments cancel out when we analyze the entire service chain.

Assume that $p_i > p_0 > v_i$ for each $i = 1, 2, \ldots, m$. The first inequality follows from the fact that the forwarder offers value-added service, e.g., customs clearance and door-to-door service, whereas the carrier does not. The second inequality follows from the observation that the direct-ship customer is typically the last-minute customer, who could not obtain cargo space in the spot market. Since $p_i > p_0$, we assume in (10) that the integrator accepts all demands for dedicated services, $\sum_{i=1}^{m} D_i$. The integrator could handle this demand using either its cargo capacity or the space elsewhere. Since $p_0 > v_i$, the integrator wants to reserve some capacity for the last-minute direct-ship customer D_0; i.e., it handles x_i units of demand D_i using its own capacity and purchases from the spot market for the excess demand $(D_i - x_i)^+$ at the rate of v_i. Specifically, the expected contribution from demand D_i to the carrier can be written as:

$$p_i D_i - v_i (D_i - x_i)^+$$
$$= p_i \min(x_i, D_i) + (p_i - v_i)(D_i - x_i)^+ \tag{11}$$
$$= (p_i - v_i) D_i + v_i \min(D_i, x_i). \tag{12}$$

In (11), the margin for the portion handled by the carrier itself is p_i, whereas that by the spot market is $p_i - v_i$. It can be seen from (12) that the larger the allotment x_i, the greater the contribution from demand D_i, but the smaller the remaining capacity for the direct-ship demand D_0 which generates higher margin (since $p_0 > v_i$). In (10), the integrator needs to determine the allotment x_i for the demand i in order to maximize the expected total contribution from both direct-ship customer and customers in need of dedicated services.

The service chain is said to be *efficient* if the total expected contribution of the chain (the integrator's expected profit) is equal to the sum of the profits of all parties. The contract which allows the efficiency to occur is said to *coordinate* the service chain [12]. The coordinating contract is desirable, since the service chain risk is shared appropriately, and there is no efficiency loss in the decentralized chain. In the analysis, we will find an equilibrium coordinating contract, if exists.

3   Analysis

For each $i = 0, 1, 2, \ldots, m$, assume that demand D_i is a nonnegative continuous random variable and is independent of one another. Let F_i be the distribution function of D_i, $\bar{F}_i$ the complementary cumulative distribution function, F_i^{-1} the quantile function, and ξ_i the density function. Define $v_i^{-1} : (0, 1) \to (0, \infty)$ as the inverse function of the utilization function; i.e., $u_i(x) = t$ if and only if $v_i^{-1}(t) = x$.

3.1   Centralized Chain

The integrator endowed with cargo capacity κ wants to choose allotments $\mathbf{x} = (x_1, x_2, \ldots, x_m)$ for demands $(D_1, D_2, \ldots, D_m)$ so that the total expected profit of the chain $\tau(\mathbf{x})$ is maximized. The integrator's problem can be formulated as the finite-horizon Markov decision process (MDP). There are m periods, and the integrator decides an allotment x_n for demand D_n in period n. As in most revenue management literature, we assume that the time periods are backward from $m, m-1, m-2, \ldots, 3, 2, 1$. In period n, let the state be the cumulative allotment usage s up to now, and let $g_n(s)$ denote the *value function*. At the beginning of the period, the integrator observes the state s and chooses an allotment x_n. The optimality equation is as follows:

$$g_n(s) = \max_{0 \leq x_n \leq \kappa - s} E[v_n \min(x_n, D_n)$$
$$+ g_{n-1}(s + \min(x_n, D_n))] \tag{13}$$

for $n = 1, 2, \ldots, m$. An allotment chosen by the integrator cannot be greater than the remaining capacity, $\kappa - s$. In period n, the integrator wants to choose an allotment x_n, which maximizes the sum of the expected contribution from demand D_n and the value function g_{n-1}, the revenue to go from period $(n-1)$ to the end of the horizon. Recall from (12) that the contribution from demand D_n is $E[(p_i - v_i)D_i + v_i \min(D_i, x_i)]$: In the optimality Eq. (13), we ignore the first term and account only for the second term, since the first term is constant and does not depend on the allotment.

The boundary conditions are as follows:

$$g_n(\kappa) = g_{n-1}(\kappa) \tag{14}$$
$$g_0(s) = E[p_0 \min(D_0, \kappa - s)]. \tag{15}$$

In (14), the value function remains constant when the entire capacity κ has been used as the allotment. In (15), the terminal value function corresponds to the expected contribution from the direct-ship customer, when the remaining capacity is

$$\kappa - s = \kappa - \sum_{n=1}^{m} \min(x_n, D_n).$$

Note that in the MDP, we assume that demands are materialized sequentially and that the cumulative allotment usage is known at the beginning of each period. Since the horizon is so short that there is no monetary discount, finding the value function given that there is no initial allotment, $g_m(0)$, is equivalent to maximizing the expected profit of the integrator, $\max\{\tau(\mathbf{x}) : \sum_{i=1}^{m} x_i \leq \kappa\}$.

The MDP formulation (13)–(15) for the integrator's problem is similar to the MDP formulation of the capacity allocation problem for the passenger airline, where the spot price v_n corresponds to the class-n fare, and the allotment x_n the protection for class-n demand; see [36] Section 2.2.2 for the multi-class capacity allocation problem for the passenger airline. If the capacity is large, directly solving the MDP may be computationally intensive, and several efficient heuristics, e.g., EMSR-A (*expected marginal seat revenue-version A*) and EMSR-B, are reviewed in Section 2.2.4 in [36].

Consider the special case when there is one forwarder. The integrator's expected profit becomes:

$$\tau(x) = \pi_1(x_1, \Omega_1) + \psi(x_1, \Omega_1)$$
$$= E[p_1 D_1 + p_0 \min(D_0, \kappa - \min(D_1, x_1))$$
$$- v_1(D_1 - x_1)^+]. \tag{16}$$

The integrator's problem of choosing an allotment x_1 in order to maximize the expected profit (16) is presented in Proposition 1.

Proposition 1

1. If $v_1 > p_0$, then $x_1^0 = \kappa$.
2. If $v_1 < p_0$, then

$$x_1^0 = \left[\kappa - F_0^{-1}\left(1 - \frac{v_1}{p_0}\right)\right]^+. \tag{17}$$

Proof. Proof can be found in Theorem 4 [2].

Recall that we assume $p_0 > v_1$. Without this assumption, the integrator's problem becomes trivial; see the first result in Proposition 1.

The optimal allotment (17) can be found using the marginal analysis. The integrator wants to find a protection for the direct-ship demand, D_0. Suppose that y units of capacity have been reserved for D_0 and that at the beginning of the season, there is a request from D_1 for an allotment. If the integrator sells now as an allotment, then it earns v_1; see (12). On the other hand, if the integrator does not sell now and reserves for the direct-ship customers, then it earns the expected margin of $p_0 P(D_0 \geq y)$. The integrator continues to protect for the direct-ship customer until $v_1 = p_0 P(D_0 \geq y)$. Re-arranging the previous term, we find that the optimal protection for the direct-ship customer is $F_0^{-1}(1-v_1/p_0)$; thus, the optimal allotment for D_1 is given in (17).

3.2   Decentralized Chain

Given that the carrier offers the contract $\Omega_i = (w_i, h_i, u_{ri})$, forwarder i chooses an optimal allotment $x_i^*(\Omega_i)$ in (6), which maximized its own expected profit.

Lemma 1. *Given the contract proposal Ω_i, the forwarder's best response allotment $x_i^*(\Omega_i)$ is as follows:*

1. *If $w_i \geq v_i$, the forwarder's expected profit is decreasing and maximized at $x_i^*(\Omega_i) = 0$.*
2. *If $w_i < v_i$ and $h_i = 0$, the forwarder's expected profit is increasing and maximized at $x_i^*(\Omega_i) = v_i^{-1}(u_{ri})$.*
3. *If $w_i < v_i$ and $h_i > 0$, the forwarder's expected profit is concave, unimodal and maximized at*

$$x_i^*(\Omega_i) = \min\{F_i^{-1}\left(1 - \frac{h_i}{v_i - w_i + h_i}\right), v_i^{-1}(u_{ri})\}. \tag{18}$$

Proof. See the proof in [2].

The first part of Lemma 1 asserts that if the discount contract price is greater than or equal to the spot price, the forwarder would not pre-book any allotment at all. On the other hand, suppose that the contract price does not exceed the spot price. If the carrier imposes no penalty cost or gives full refund for the unused allotment (i.e., $h_i = 0$), the forwarder's expected profit is increasing, and the forwarder would choose the largest allotment that satisfies the required allotment utilization. The last part asserts that if there is a positive penalty cost or a partial refund is given (i.e., $h_i > 0$), the forwarder should pre-book the allotment, which balances the cost associated with the unused allotment and the opportunity cost from not having enough allotment. An optimal allotment in Lemma 1 bears a striking resemblance to the optimal order quantity in the newsvendor (single-period) inventory model. In the newsvendor model, an optimal order quantity is chosen such that the expected total cost $E[c_u(D-q)^+ + c_o(q-D)^+]$ is minimized: $q^* = F^{-1}(1 - c_o/(c_u + c_o))$ where c_u (resp., c_o) is the unit *underage* (resp., *overage*) cost from ordering less (resp., more) than demand, and F is the distribution of demand D; see a standard textbook in operations management for the newsvendor model, e.g., chapter 5 in [29]. In ours, the overage corresponds to the penalty cost for the unused allotment h_i, and the underage is the saving forgone $v_i - w_i$ if the forwarder purchases from the spot market instead of using the allotment. The critical ratio $1 - c_o/(c_u + c_o)$ becomes as in (18).

Let $\mathbf{x}^0 = (x_i^0; i = 1, 2, \ldots, m)$ be the integrator's optimal allotment, which maximizes the expected profit of the centralized chain. Recall that $\mathbf{x}^*(\mathbf{H}) = (x_i^*(\Omega_i); i = 1, 2, \ldots, m)$ denotes the optimal allotments chosen by the forwarders, in order to maximizes their own profits. Below, we will determine the equilibrium of the game, in which each party maximizes its own expected profit. For shorthand, denote $\Lambda_i = 1/F_i(x_i^0) - 1$.

Theorem 1. *Suppose that $\bar{u}_{ri} = 0$ for each $i = 1, 2, \ldots, m$ and that there exists $0 < \bar{h}_i < v_i \Lambda_i$ such that $\pi_i(x_i^0, \bar{\Omega}_i) = \epsilon_i$ where*

$$\bar{w}_i = v_i - \bar{h}_i/\Lambda_i \tag{19}$$

Then, the contract $\bar{\mathbf{H}} = (\bar{\Omega}_i; i = 1, 2, \ldots, m)$ where $\bar{\Omega}_i = (\bar{w}_i, \bar{h}_i, \bar{u}_{ri})$ is an equilibrium of the game and coordinates the service chain.

Proof. In the decentralized chain, the carrier wants to find $\mathbf{H}$ which maximizes

$$\psi(\mathbf{x}^*(\mathbf{H}), \mathbf{H}) = \tau(\mathbf{x}^*(\mathbf{H})) - \sum_{i=1}^{m} \pi_i(x_i^*(\Omega_i), \Omega_i) \tag{20}$$

$$\text{subject to: } \pi_i(x_i^*(\Omega_i), \Omega_i) \geq \epsilon_i; i = 1, 2, \ldots, m. \tag{21}$$

The Karush-Kuhn-Tucker (KKT) conditions that are necessary for a point $\bar{\mathbf{H}} = ((\bar{w}_i, \bar{h}_i, 0); i = 1, 2, \ldots, m)$ to solve (20)–(21) are as follows: There exists a multiplier $\bar{\lambda}_i \geq 0$ for $i = 1, 2, \ldots, m$ satisfying

$$\frac{\partial \psi}{\partial \bar{w}_j} - \sum_{i=1}^{m} \bar{\lambda}_i \frac{\partial \pi_i}{\partial \bar{w}_j} = 0; \quad j = 1, 2, \ldots, m \tag{22}$$

$$\frac{\partial \psi}{\partial \bar{h}_j} - \sum_{i=1}^{m} \bar{\lambda}_i \frac{\partial \pi_i}{\partial \bar{h}_j} = 0; \quad j = 1, 2, \ldots, m \tag{23}$$

$$\bar{\lambda}_j (\epsilon_j - \pi_j) = 0; \quad j = 1, 2, \ldots, m. \tag{24}$$

Suppose that $\bar{\lambda}_j > 0$. It follows from (24) that

$$\pi_j(x_j^*(\bar{\Omega}_j), \bar{\Omega}_j) = \epsilon_j.$$

for each $j = 1, 2, \ldots, m$. The contract parameters are chosen such that the forwarder earns exactly its reservation profit. Recall that $\psi = \tau - \sum_i \pi_i$. Note that

$$\frac{\partial \psi}{\partial \bar{w}_j} = \frac{\partial \tau(\mathbf{x}^*(\bar{\mathbf{H}}))}{\partial \bar{w}_j} - \frac{\partial \pi_j(x^*(\bar{\Omega}_i), \bar{\Omega}_i)}{\partial \bar{w}_j}.$$

Equation (22) becomes

$$\frac{\partial \tau(\mathbf{x}^*(\bar{\mathbf{H}}))}{\partial \bar{w}_j} - (1 + \bar{\lambda}_j) \frac{\partial \pi_j(x_j^*(\bar{\Omega}_j), \bar{\Omega}_j)}{\partial \bar{w}_j} = 0. \tag{25}$$

The contract parameter (19) is chosen such that

$$x_j^*(\bar{\Omega}_j) = x_j^0 = F_i^{-1}\left(1 - \frac{\bar{h}_j}{v_i - \bar{w}_j + \bar{h}_j}\right). \tag{26}$$

(In other words, after terms are re-arranged, (26) becomes (19).) It follows from (26) and Lemma 1 that the forwarders' best responses also maximize the integrator's expected profit: The necessary conditions are as follows:

$$\frac{\partial \tau}{\partial x_j} = \frac{\partial \pi_j}{\partial x_j} = 0 \quad \text{for } j = 1, 2, \ldots, m. \tag{27}$$

Applying the (multivariable) chain rule to (25) and invoking the necessary conditions (27), we conclude that (22) holds. Similarly, we can show that (23) holds. The KKT conditions are satisfied.

At the point $\bar{\mathbf{H}} = ((\bar{w}_i, \bar{h}_i, 0) : i = 1, 2, \ldots, m)$, the carrier receives the maximum chain's expected profit less the total reservation profits of all forwarders:

$$\psi(\mathbf{x}^0, \bar{\mathbf{H}}) = \tau(\mathbf{x}^0) - \sum_{i=1}^{m} \epsilon_i.$$

The carrier cannot do better than this; thus, the point $\bar{\mathbf{H}}$ is an equilibrium solution to our sequential game.

At the equilibrium, the carrier offers a contract parameter $\bar{\mathbf{H}}$ so that the forwarder's best response is equal to the integrator's optimal allotment. The forwarder earns exactly its reservation profit. In this sequential game, the carrier is the leader, and there is a so-called leader's *first-mover advantage*.

In Theorem 1, the carrier imposes a strictly positive penalty cost for the unused portion of the allotment. Theorem 2 presents an equilibrium coordinating contract, in which no penalty cost is imposed.

Theorem 2. *Suppose that $\bar{h}_i = 0$ for each $i = 1, 2, \ldots, m$ and that there exists $0 < \bar{w}_i < v_i$ such that $\pi_i(x_i^0, \bar{\Omega}_i) = \epsilon_i$ where*

$$\bar{u}_{ri} = u_i(x_i^0). \tag{28}$$

Then, the contract $\bar{\mathbf{H}} = (\bar{\Omega}_i; i = 1, 2, \ldots, m)$ where $\bar{\Omega}_i = (\bar{w}_i, \bar{h}_i, \bar{u}_{ri})$ is an equilibrium of the game and coordinates the service chain.

Proof. Recall from Lemma 1 that if $\bar{w}_i < v_i$, then $x_i^*(\bar{\Omega}_i) = v_i^{-1}(\bar{u}_{ri})$. From the assumption that D_i is continuous, we have that v_i^{-1} is a one-to-one function and conclude that $x_i^*(\bar{\Omega}_i) = x_i^0$. The rest of the proof is similar to that in Theorem 1.

Theorem 2 states that if the penalty cost for the unused allotment is zero and that the contract price is less than the spot price, then the forwarder chooses the largest allotment which satisfies the required utilization. In the carrier's problem, the required utilization becomes the decision variable. Setting the required utilization equal to the expected utilization evaluated at the integrator's optimal allotment, the chain becomes efficient.

In practice, it is uncommon to a full-refund contract (i.e., $h_i = 0$). To ensure its high customer service level, the forwarder may ask for a very large allotment (much greater than its anticipated customer demand) and release the unwanted allotment so late that the carrier might not have enough time to re-sell to direct-ship customers. To prevent the forwarder to pre-book a large allotment, Theorem 2 suggests that the carrier needs to impose the minimum utilization requirement.

In Theorems 1 and 2, we present two contract schemes that can coordinate the service chain, and the optimal contract has two parameters. The discount contract price is included in the two optimal schemes. In Theorem 1, the other contract parameter is the positive penalty cost, whereas in Theorem 2, the other parameter is the required allotment utilization. We do not need to have a three-parameter contract in order for the air-cargo service chain to be efficient. Our result provides some managerial insights that can help the carrier to negotiate the contractual terms with the forwarders.

4   Numerical Examples

We provide a numerical example to illustrate our approach of finding an equilibrium in the air-cargo contract game. As an illustration, assume that there are $m = 2$ forwarders and that demands are independent. For each $i = 0, 1, 2$, demand D_i (in kilogram) follows the gamma distribution with the shape parameter α_i and the rate parameter β_i, shown in Table 1. The carrier's cargo capacity is assumed to be $\kappa = 800$ kg. The per-unit price (in THB/kilogram) the forwarder charges its customer, the per-unit price the carrier charges its direct-ship customer and the spot prices observed by the forwarders are also given in Table 1. These parameters are similar to those on the Bangkok-Dublin (BKK-DUB) route in the medium season in 2014.

Table 1. Parameters for the numerical example.

	p_i	v_i	α_i	β_i
AC ($i = 0$)	60.00	N/A	2.30	0.01
FF1 ($i = 1$)	69.00	55.00	5.10	0.01
FF2 ($i = 2$)	62.00	50.00	3.60	0.01

Calculation details are provided below. Recall the forwarder i's expected profit (1)

$$E[p_i D_i - w_i \min(D_i, x_i) - v_i(D_i - x_i)^+ - h_i(x_i - D_i)^+]. \tag{29}$$

For the first term in (29), the forwarder's mean demand $E[D_i]$ is the expected value of the gamma distribution:

$$E[D_i] = \alpha_i/\beta_i.$$

For the second term in (29), we evaluate the expected allotment usage $E[\min(D_i, x)]$ using the limited expected value (LEV) function. The LEV function for gamma random variable, Y, with the shape parameter α and the scale parameter θ is

$$E[\min(Y, x)] = \alpha\theta\Gamma(\alpha + 1; x/\theta) + x[1 - \Gamma(\alpha; x/\theta]$$

where $\Gamma(\alpha; x)$ is the incomplete gamma function defined by

$$\Gamma(\alpha; x) = \frac{1}{\Gamma(\alpha)} \int_0^x t^{\alpha-1} e^{-t} dt$$

and $\Gamma(\alpha)$ is the gamma function, defined by

$$\Gamma(\alpha) = \int_0^\infty x^{\alpha-1} e^{-x} dx.$$

(For the gamma distribution, the scale parameter is equal to the reciprocal of the rate parameter.) For the last two terms in (29), we use the following

$$E[(D_i - x)^+] = E[D_i] - E[\min(D_i, x)]$$
$$E[(x - D_i)^+] = x - E[\min(D_i, x)]$$

where the expected allotment usage is calculated previously using the LEV function. Recall the carrier's expected profit (7):

$$E[p_0 \min(D_0, \kappa - \sum_{i=1}^{m} \min(D_i, x_i)) + \sum_{i=1}^{m} t_i(x_i, D_i)].$$

To calculate the first term, we again make use of the LEV.

$$E[\min(D_0, \kappa - \sum_{i=1}^{m} \min(D_i, x_i)]$$

$$= \int_0^{x_m} \cdots \int_0^{x_1} E[\min(D_0, \kappa - \sum_{i=1}^{m} t_i)] \Pi_{i=1}^{m} dF_i(t_i)$$

$$+ \int_{x_m}^{\infty} \cdots \int_{x_1}^{\infty} E[\min(D_0, \kappa - \sum_{i=1}^{m} x_i)] \Pi_{i=1}^{m} dF_i(t_i)$$

$$= \int_0^{x_m} \cdots \int_0^{x_1} E[\min(D_0, \kappa - \sum_{i=1}^{m} t_i)] \Pi_{i=1}^{m} dF_i(t_i)$$

$$+ E[\min(D_0, \kappa - \sum_{i=1}^{m} x_i)] (\Pi_{i=1}^{m} \bar{F}_i(x_i)).$$

The integrator's expected profit (10) can be found similarly. Throughout our numerical examples, calculations are done in R. For instance, the cubature package is used for numerical integration over simplexes, qgamma, dgamma and pgamma return the quantile, density and cumulative distribution functions of the gamma distribution.

4.1   Centralized Chain

Suppose that we use the variant of EMSR algorithm given in Section 5.2 [22]. The allotment x_i for demand D_i is given as follows:

$$x_i^0 = F_i^{-1}(1 - v_{i+1}/v_i); \quad i = 0, 1, 2, \ldots, m - 1 \tag{30}$$

$$x_m^0 = (\kappa - \sum_{i=0}^{m-1} x_i^0)^+ \tag{31}$$

where $v_0 = p_0$ is the per-unit price to the direct-ship customer. Using (30)–(31), we obtain that

$$\mathbf{x}^0 = (x_0^0, x_1^0, x_2^0) = (63, 243, 494)$$

and the integrator's expected profit is $\tau(\mathbf{x}^0) = 42458$. The integrator would allocate $x_1^0 = 243$ kg to demand D_1 and $x_2^0 = 494$ to D_2. The remaining capacity of $x_0^0 = 63$ is reserved for the direct-ship customer. The expected utilizations by demands D_1 and D_2 are $u_1(x_1^0) = 97.99\%$ and $u_x(x_2^0) = 66.30\%$, respectively.

4.2   Decentralized Chain

Assume that the reservation profits are $\epsilon_1 = 12000$ and $\epsilon_2 = 10000$. (Note that these two values are greater than the lower bounds, $E[(p_1 - v_1)D_1] = 7140$ and $E[(p_2 - v_2)D_2] = 4320$, when the forwarders use only the spot markets.) We illustrate how to find an equilibrium coordinating contract using Theorem 1 in Example 1 and Theorem 2 in Example 2

Example 1. Consider the contract C1: The carrier offers the contract price $w_1 = 49.50$ (resp., $w_2 = 49.00$) and the penalty cost $h_1 = 55.00$ (resp., $h_2 = 1.33$) to forwarder 1 (resp., 2). Note that the penalty cost is chosen such that $h_i = \lambda_i v_i \Lambda_i$ where $\lambda_i \in (0, 1)$ is given in Table 2, and that the contract price w_i is given by (19). (We want $\lambda_i < 1$ since we are trying to find $\bar{h}_i < v_i \Lambda_i$; see Theorem 1.) If forwarder i were to accept the contract proposal, it would choose $x_i^*(\Omega_i) = x_i^0$. Nevertheless, both forwarders reject this contract C1, because their expected profits given the contract parameters are less than their reservation profits. On the other hand, both forwarders accept contract C2 in Table 2, since their expected profits are greater than their reservation profits. Note that in contract C2 when the larger discount is given to the forwarder (i.e., w_1 decreases from 49.50 to 27.50), the penalty cost for the unused portion of the allotment also becomes larger (i.e., h_1 increases from 55 to 275). Both contracts C1 and C2 coordinate the chain; i.e., $x_i^*(\Omega_i) = x_i^0$, and the total profit in the chain is maximized and equal the optimal integrator's profit, Contract C1 is rejected by the forwarders, whereas C2 is accepted: The expected profits of forwarders 1 and 2 are 12342 and 11407, respectively, and the carrier's expected profit is 18709, which is about 44% of the chain's optimal profit. Contract C2 is not an equilibrium solution in our sequential game, since the carrier can still improve its profit while giving the forwarders at least their reservation profits.

Theorem 1 states that at the equilibrium, the forwarders earn exactly their reservation profits. Note that

$$(\pi_1, \pi_2) \text{ given C1 } \leq (\epsilon_1, \epsilon_2) \leq (\pi_1, \pi_2) \text{ given C2.}$$

We can use a bisection method to search for an optimal contract with (λ_1, λ_2)

$$(0.10, 0.10) \leq (\lambda_1, \lambda_2) \leq (0.50, 0.50)$$

such that the forwarders' expected profits are equal to their reservation profits. If we stop when $|\pi_i - \epsilon_i| \leq \delta_i$ where δ_i is a pre-specified tolerance, say 100 THB, then we obtain contract C3 in Table 2. The discount contract price for forwarder 1 (resp., 2) is 28.88 THB/kg (resp., 29.75 THB/kg), and the penalty cost for the unused portion of the allotment is 261.25 THB/kg (resp., 5.38 THB/kg). The forwarders earn (approximately) their reservation profits, and the carrier earns the rest, about of 48% of the optimal integrator's expected profit.

At the equilibrium contract with strictly positive penalty cost, the carrier does not need to impose the minimum utilization requirement. At the equilibrium, the two-parameter contract is sufficient to coordinate the chain.

The two contract parameters (w_i, h_i) can be interpreted differently using (3)–(4). For forwarder 1, $\bar{\Omega}_1 = (\bar{w}_1, \bar{h}_1) = (28.88, 261.55)$, the wholesale price for the allotment $x_1^*(\bar{\Omega}_1) = 243$ is 28.88, and the payment upfront is 7014.90; after demand D_1 materializes, the unused allotment is charged at the penalty rate of 232.38. For forwarder 2, $\bar{\Omega}_2 = (\bar{w}_2, \bar{h}_2) = (29.75, 5.38)$, the wholesale price for the allotment $x_2^*(\bar{\Omega}_1) = 494$ is 29.75, and the payment upfront is 14704.32; after demand D_2 materializes, the unused allotment is returned at the refund rate of 24.37.

Table 2. Possible contracts with strictly positive penalty costs $h_i > 0$.

Contract	C1	C2	C3
λ_1	0.10	0.50	0.48
λ_2	0.10	0.50	0.41
h_1	55.00	275.00	261.25
h_2	1.33	6.64	5.38
w_1	49.50	27.50	28.88
w_2	45.00	25.00	29.75
$x_1^*(\Omega_1)$	243	243	243
$x_2^*(\Omega_2)$	494	494	494
$\pi_1(x_1^*, \Omega_1)$	8180	12342	12082
$\pi_2(x_2^*, \Omega_2)$	5737	11407	10060
$\psi(\mathbf{x}^*, \mathbf{H})$	28541	18709	20316
Total expected profit	42458	42458	42458

Example 2. The equilibrium coordinating contract found in Theorem 1 corresponds to the positive penalty cost (or partial refund payment) for the unused allotment. In Example 2, we illustrate how to use Theorem 2 to find an equilibrium coordinating contract with no penalty cost (or full refund payment), $h_i = 0$ for $i = 1, 2$. By solving for $\pi_i(x_i^0, (\bar{w}_i, 0, \bar{u}_{ri})) = \epsilon_i$ as specified in Theorem 2, we find the contract price as follows:

$$\bar{w}_i = \frac{E[p_i D_i - v_i(D_i - x_i^0)^+] - \epsilon_i}{E[\min(D_i, x_i^0)]}. \tag{32}$$

Substituting all input parameters and the integrator's optimal allotments into (32), we obtain the equilibrium coordinating contract C4 in Table 3. With contract C4, the carrier imposes no penalty cost on the unused allotment but needs to impose the required allotment utilization of 97.99% and 66.30% for forwarders 1 and 2, respectively. Also, observe that when the carrier imposes no penalty cost, the discount terms in C4 are not as generous as those in C3.

Table 3. Equilibrium coordinating contracts with two parameters.

Forwarder	C3		C4	
	1	2	1	2
Disc. price	28.88	29.75	34.58	32.67
Penalty cost	261.55	5.38	N/A	N/A
Req. Util. (%)	N/A	N/A	99.79	66.30

Our numerical examples illustrate how to construct equilibrium coordinating contracts. Using Theorem 1, we construct a two-parameter contract C3 with the discount contract price and the positive penalty cost. Using Theorem 2, we construct another two-parameter contract C4 with the discount contract price and the minimum allotment utilization requirement.

5   Concluding Remark

We consider the air-cargo service chain, which consists of the carrier and multiple freight forwarders. Each of the forwarders and the carrier may enter into an allotment contract before the selling season starts. We formulate the contract design problem as the Stackelberg game, in which the carrier is the leader and proposes the contract. The proposed contract in this article has three parameters, namely the discount contract price, the penalty cost for the unused portion of the allotment and the required allotment utilization. Each of the forwarders responses by choosing the best allotment, which maximizes its expected profit, given the carrier's offered contract. We show that the two-parameter contract is sufficient to coordinate the chain. At the equilibrium, the forwarders' optimal allotments correspond to the integrator's optimal allotments that maximize the total expected profit in the chain, and the forwarders receive exactly their reservation profits.

A few extensions are as follows: When forwarders' services are similar and substitutable, the demand to a particular forwarder depends on both its price and the competitors' prices. Along the same lines, the carrier's direct-ship demand may depend on the carrier's direct-ship price and the forwarders' prices. We can extend ours to include the price competition. For instance, suppose that the pricing decision is made before the contract process begins. Then, the equilibrium prices and the corresponding demands become our input parameters (i.e., p_i and D_i for $i = 0, 1, 2, \ldots, m$) in this article. Another extension is to capture asymmetric information between the carrier and the forwarder. The forwarder may possess some private information, e.g., its customer demand, its spot price and its reservation profit. We can study how to design an optimal contract. For instance, the carrier may offer a menu of possible contracts, and the forwarder optimally selects from the menu. Finally, we can apply our approach to other RM industry, in which a portion of the perishable capacity is sold through a medium- or long-term contract. For instance, in the passenger airline industry, the airline usually blocks a pre-specified number of seats to a wholesaler/agent or other airlines under the interline or codeshare agreements, which are agreed upon prior to the start of the selling season. We hope to pursue these or related problems.

Acknowledgements. This research was supported in part by the Thailand Research Fund and Office of the Higher Education Commission, Thailand (Grant RSA58-Kannapha-Amaruchkul). The views expressed in this paper are that of the author and do not necessarily reflect the views of the Thailand Research Fund and Office of the Higher Education Commission, Thailand. Data collection and in-depth interviews were partially conducted by the Ms. Apinya Theppanomrat, one of the full-time master students in the logistics management (LM) program at NIDA, and Ms. Narisara Lueang-orn, one of the part-time master LM students, who has been working at the freight forwarder company (in the numerical example) for several years.

References

1. Accenture: 2015 air cargo survey: taking off for higher profitability (2015). www.accenture.com. Accessed 3 July 2016
2. Amaruchkul, K.: Game-theoretic analysis of air-cargo allotment contract. In: International Conference on Operations Research and Enterprise Systems (ICORES 2018), Funchal, Portugal (2018)
3. Amaruchkul, K., Cooper, W., Gupta, D.: Single-leg air-cargo revenue management. Transp. Sci. **41**(4), 457–469 (2007)
4. Amaruchkul, K., Cooper, W., Gupta, D.: A note on air-cargo capacity contracts. Prod. Oper. Manage. **20**(1), 152–162 (2011)
5. Amaruchkul, K., Lorchirachoonkul, V.: Air-cargo capacity allocation for multiple freight forwarders. Transp. Res. Part E **47**(1), 30–40 (2011)
6. Barz, C.: Risk-Averse Capacity Control in Revenue Management. Springer, Heidelberg (2007). https://doi.org/10.1007/978-3-540-73014-9
7. Barz, C., Gartner, D.: Air cargo network revenue management. Transp. Sci. **50**(4), 1206–1222 (2016)
8. Bazaraa, M., et al.: The Asia Pacific air cargo system. Research Paper No: TLI-AP/00/01. The Logistics Institute-Asia Pacific. National University of Singapore and Georgia Institute of Technology (2001). http://www.tliap.nus.edu.sg/Library/default.aspx. Accessed 06 Jan 2007
9. Becker, B., Dill, N.: Managing the complexity of air cargo revenue management. J. Revenue Pricing Manage. **6**(3), 175–187 (2007)
10. Billings, J., Diener, A., Yuen, B.: Cargo revenue optimisation. J. Revenue Pricing Manage. **2**(1), 69–79 (2003)
11. Boeing Company: World air cargo forecast 2016–2017 (2016). http://www.boeing.com/commercial/market/cargo-forecast/. Accessed 13 Apr 2018
12. Cachon, G.: Supply chain coordination with contracts. In: de Kok, A., Graves, S. (eds.) Handbooks in Operations Research and Management Science: Supply Chain Management. Elsevier, Amsterdam (2003)
13. Chiang, W., Chen, J., Xu, X.: An overview of research on revenue management: current issues and future research. Int. J. Revenue Manage. **1**(1), 97–128 (2007)
14. DeLain, L., O'Meara, E.: Building a business case for revenue management. J. Revenue Pricing Manage. **2**(4), 368–377 (2004)

15. Feng, B., Li, Y., Shen, H.: Tying mechanism for airlines' air cargo capacity allocation. Eur. J. Oper. Res. **244**, 322–330 (2015)
16. Feng, B., Li, Y., Shen, Z.: Air cargo operations: literature review and comparison with practices. Transp. Res. Part C **56**, 263–280 (2015)
17. Gupta, D.: Flexible carrier-forwarder contracts for air cargo business. J. Revenue Pricing Manage. **7**(4), 341–356 (2008)
18. Hellermann, R.: Options contracts with overbooking in the air cargo industry. Decis. Sci. **44**(2), 297–327 (2013)
19. Hendricks, G., Elliott, T.: Implementing revenue management techniques in an air cargo environment. (2010). www.unisys.com/transportation/insights. Accessed 3 Mar 2010
20. Hoffmann, R.: Dynamic capacity control in cargo revenue management-a new heuritic for solving the single-leg problem efficiently. J. Revenue Pricing Manage. **12**(1), 46–59 (2013)
21. Ingold, A., McMahon-Beattie, U., Yeoman, I.: Yield Management: strategies for the service industries. South-Western Cengage Learning, Bedford Row, London (2000)
22. International Air Transport Association: Airline Revenue Management: Course eTextbook. International Aviation Training Program, Montreal, Quebec (2012)
23. International Air Transport Association: IATA cargo strategy (2016). www.iata.org. Accessed 3 July 2016
24. Kasilingam, R.: Air cargo revenue management: characteristics and complexities. Eur. J. Oper. Res. **96**(1), 36–44 (1996)
25. Levin, Y., Nediak, M., Topaloglu, H.: Cargo capacity management with allotments and spot market demand. Oper. Res. **60**(2), 351–365 (2012)
26. Lin, D., Lee, C., Yang, J.: Air cargo revenue management under buy-back policy. J. Air Transp. Manage. **61**, 53–63 (2017)
27. McGill, J., van Ryzin, G.: Revenue management: research overview and prospects. Transp. Sci. **33**(2), 233–256 (1999)
28. Moussawi-Haidar, L.: Optimal solution for a cargo revenue management problem with allotment and spot arrivals. Transp. Res. Part E **72**, 173–191 (2014)
29. Nahmias, S.: Production and Operations Research. McGraw-Hill Inc., New York (2009)
30. Netessine, S., Shumsky, R.: Introduction to the theory and practice of yield management. INFORMS Trans. Educ. **3**(1), 34–44 (2002)
31. Phillips, R.: Pricing and Revenue Optimization. Stanford University Press, Stanford (2005)
32. Prior, R., Slavens, R., Trimarco, J.: Menlo worldwide forwarding optimizes its network routing. Interfaces **34**(1), 26–38 (2004)
33. Sales, M.: The Air Logistics Handbook: Air Freight and the Global Supply Chain. Routledge, New York (2013)
34. Sigworth, D.: Contracting: making the rate case. Traffic World **268**(17), 32–33 (2004)
35. Slager, B., Kapteijns, L.: Implementation of cargo revenue management at KLM. J. Revenue Pricing Manage. **3**(1), 80–90 (2004)
36. Talluri, K., van Ryzin, G.: The Theory and Practice of Revenue Management. Kluwer Academic Publishers, Boston (2004)
37. Tang, C.: A scenario decomposition-genetic algorithm method for solving stochastic air cargo container loading problems. Transp. Res. Part E **45**(6), 725–739 (2011)

38. Tao, Y., Chew, E., Lee, L., Wang, L.: A capacity pricing and reservation problem under option contract in the air cargo freight industry. Comput. Ind. Eng. **110**, 560–572 (2017)
39. Wada, M., Delgado, F., Pagnoncelli, B.: A risk averse approach to the capacity allocation problem in the airline cargo industry. J. Oper. Res. Soc. **68**(6), 643–651 (2017)
40. Woods, R.: Top 5 perishables with the fastest-growing demand for air cargo (2017). https://aircargoworld.com/allposts/top-5-perishables-with-the-fastest-growing-demand-for-air-cargo/. Accessed 13 Apr 2018
41. Wu, Y.: Modelling of containerized air cargo forwarding problems under uncertainty. J. Oper. Res. Soc. **62**(7), 1211–1226 (2011)
42. Yang, S., Chen, S., Chen, C.: Air cargo fleet routing and timetable setting with multiple on-time demands. Transp. Res. Part E: Logist. Transp. Rev. **42**(5), 409–430 (2006)
43. Yeoman, I., McMahon-Beattie, U.: Revenue Management and Pricing: Case Studies and. Thomson Learning, Bedford Row (2004)
44. Yeung, J., He, W.: Shipment planning, capacity contracting and revenue management in the air cargo industry: a literature review. In: Proceedings of the 2012 International Conference on Industrial Engineering and Operations Management, Istanbul, July 2012
45. Zhang, C., Xie, F., Huang, K., Wu, T., Liang, Z.: MIP models and a hybrid method for the capacitated air-cargo network planning and scheduling problems. Transp. Res. Part E **103**, 158–173 (2017)

Computational Study of Emergency Service System Reengineering Under Generalized Disutility

Marek Kvet[✉], Jaroslav Janáček, and Michal Kvet

Faculty of Management Science and Informatics,
University of Žilina, Univerzitná 8215/1, 010 26 Žilina, Slovakia
{marek.kvet, jaroslav.janacek,
michal.kvet}@fri.uniza.sk

Abstract. Emergency medical service system structure is determined by deployment of limited number of the service providing centers. The objective of the designer is to minimize the total discomfort of all system users. Thus, the problem often takes the form of the weighted p-median problem. Since population and demands for service change in time and space, current service center deployment may not meet the requirements of the users and service providers neither. We suggest and discuss a mathematical model for system reengineering under the generalized disutility. Formulation of the generalized disutility follows from the idea that the individual user's disutility is caused by positions of more than one located service center. Generalized disutility enables to model the system performance more realistically. It enables to take into account also such situations in which the nearest service center may be temporarily unavailable due to satisfying another demand. This approach represents an extension of our previous research, in which only the nearest center was taken as a source of individual user's demand satisfaction.

Keywords: Emergency medical service · System reengineering ·
Generalized disutility · Radial formulation

1 Introduction

An emergency service system as the medical emergency system, system of fire brigades and system of police stations are designed for given geographical area to satisfy the demand of population living in the area for more secure life. Associate service is provided from a given number of service centers and the overwhelming objective used for the design evaluation is the average time necessary to deliver service from a service center to the user location, in which a demand for service has occurred.

Host of models assume that serviced population is concentrated to a finite number of dwelling places of the considered area. Frequency of the demand occurrence at a given place is proportional to the number of inhabitants of the given town or village. A finite set of possible service center locations is assumed and also the assumption is made that a user demand is serviced from the nearest located service center. This way, the weighted p-median problem formulation is used to the emergency service system

© Springer Nature Switzerland AG 2019
G. H. Parlier et al. (Eds.): ICORES 2018, CCIS 966, pp. 198–219, 2019.
https://doi.org/10.1007/978-3-030-16035-7_11

design and solving the underlying problem to optimality [2, 3, 6, 10]. The original way of modelling was based on the location-allocation decision variables and constraints [2], where an occurring demand is assigned to exactly one possible center location. As concerns usage of a general IP-solver, the size of the solved integer programming problem must be taken into account. In real problems, the number of serviced users takes the value of several thousands, and the number of possible service center locations can take this value as well [1]. The number of possible service center locations seriously impacts the computational time and the memory of computer due to used branch-and-bound method, which stores the unfathomed nodes of the inspected searching tree for the further processing. Therefore, the direct attempt at solving the problem described by a location-allocation model often fails, when larger instances are solved by a commercial IP-solver. Then, another approach using so-called radial formulation was developed to avoid the particular assignment of user's locations to the located service centers. The radial approach successfulness is based on the fact that there is only finite set of radii, which must be taken into account [4, 5, 7]. To accelerate the p-median problem solving process performed by commercial IP-solvers, an approximate approach has been developed [8]. This approach uses an approximation of a common time distance between a service center location and a user by some predetermined time distances and gives near to optimal results in the case of integer time distances. Accuracy of the resulting solution can be arbitrarily improved by usage of smaller units for the time-distance evaluation.

A bit different situation occurs, when reengineering of a current emergency service system is performed. The necessity of system updating usually follows from the fact that distribution of demands for service has been developing in time and space and thus, the originally determined center locations do not suit both serviced public and providers operating the service centers. Contrary to the original system design, the current service providers suggest changes in the center deployment and their suggestion may be in a conflict with public interests. That is why the system administrator permits system reengineering only subject to some formal rules, which are intended to prevent worsening the service accessibility. The considered formal rules are quantified by a maximal number of provider's centers, which are allowed to change their locations and by the maximal distance between a current center location and a possible new location. Generally, addition of constraints may significantly spoil the computational time necessary to obtain the optimal solution of the problem. The study [12] showed, that they do not impact the computational time, when a user demand is serviced from the nearest located center.

In this paper, we deal with more general model of the emergency medical system design under reengineering. We assume that service of a user demand is provided from the nearest center only in the case, when the center is not occupied by servicing a former demand. Otherwise, the user's demand is serviced from the nearest unoccupied center. Initial emergency system design considering the failing centers was studied in [14] and the associated radial formulation was presented in [11]. Nevertheless, the reengineering of service system with failing centers has not been studied yet. Therefore, we focus on the influence of the formal rule constraints on best possible service availability in the service system and on the associated computational process convergence.

We provide a reader with a radial model of emergency service system reengineering with failing centers under rules imposed by the system administrator. We perform a computational study to find whether real-sized instances of the problem are solvable using a common IP-solver.

The remainder of the paper is organized as follows. The next section is devoted to the radial model formulation, in which temporarily failing centers are considered. In Sect. 3, administrator's auxiliary rules are introduced and various ways of their implementation in the associated models are discussed. Section 4 consists of numerical experiments focused on three goals. The first one is connected with optimization of model parameters, which influence both model size and the result accuracy. The second goal consists in answering the question: How do the individual parameters of the administrator's rules influence the resulting average response time of the emergency service? The third portion of the numerical experiments aims at investigation of mutual impact of the observed formal rules parameters. The conclusion summarizes obtained findings and contains possible directions of a further research.

2   Reengineering of a Service System with Failing Centers

The emergency system reengineering was originally studied in [12], where the radial model of the problem was introduced. The basic idea follows from the analysis of current service center deployment, which may not be optimal due to changing demands and development of the underlying transportation network. To explain the problem in more details, consider the simple example depicted in Fig. 1. We assume that the left graph represents current deployment of four service centers marked by blue color. All the vertices represent possible demand points. To evaluate the current center deployment, the sum of distances from each network node to the nearest located center was used as the quality criterion. Here, it takes the value of 66. If we allowed changes in current service center locations and moved a service center from the node 2 to the node 6, we would perform system reengineering and we could achieve better value of the criterion. The new system design is depicted on the right graph and its evaluation is 64. By this small example we demonstrated the principle and goal of system reengineering.

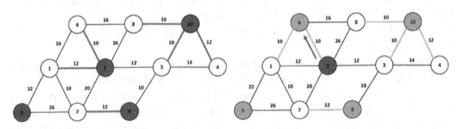

Fig. 1. Simple example of emergency service system reengineering [12].

In this paper, we study the reengineering problem under generalized disutility with the assumption that the service is provided by more than one provider. We also extend the research published in [13]. To describe the problem of the users' disutility minimization by changing the deployment of centers belonging to one considered provider, we introduce several necessary notations. Let us denote J as a finite set of all users (dwelling places), where b_j denotes a volume of expected demand of user $j \in J$. Let I be a finite set of all possible center locations. We use the symbol d_{ij} to denote the integer network time distance between locations i and j, where $i, j \in I \cup J$. The maximal considered distance is m. The current emergency service center deployment is described by union of two disjoint sets of located centers I_L and I_F, where I_L contains p centers of the considered provider and I_F is the set of centers belonging to the other providers. The considered provider performs updating of his part of the system by changing locations of the centers from I_L. The center locations from I_L can be relocated within the set $I_R = I - I_F$. Locations of centers from I_F stay unchanged.

Let value q_k represent probability of the case that the $k-1$ nearest centers are occupied and the k-th nearest center is the first one, which is available [9, 14].

The probabilities q_k for $k = 1, \ldots, r$ are positive real values, which meet the following inequalities $q_1 \geq q_2 \geq \ldots \geq q_r$ and depend only on the order of distances from the user to the r nearest centers. In this paper, the generalized disutility perceived by a user is modelled by a sum of weighted time distances from the r nearest located centers. Mentioned concept of generalized disutility is depicted in the following Fig. 2.

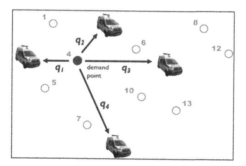

Fig. 2. Concept of generalized disutility, where the distances to the service centers are weighted by probability coefficients q_k. The shortest distance is weighted by the highest coefficient.

To complete the following radial model, we introduce coefficients a_{ij}^s for each pair i, j of possible center location and location of the user j. We define $a_{ij}^s = 1$ if and only if $d_{ij} \leq s$ and $a_{ij}^s = 0$ otherwise for each $s = 0, 1, \ldots, m-1$.

To describe decisions on new center deployment, we introduce location zero-one variables y_i defined for each $i \in I_R$. The variable y_i takes the value of one, if a service center is to be located at i and it takes the value of zero otherwise. To express the total distance necessary for user demand satisfaction in the radial manner, we introduce auxiliary zero-one variables x_{jsk} for $j \in J, s \in 0 \ldots m-1, k \in 1 \ldots r$ to model the disutility contribution value of the k-th nearest service center to the user j. The variable

x_{jsk} takes the value of 1 if the k-th smallest disutility contribution for the user $j \in J$ is greater than s and it takes the value of 0 otherwise. Then the expression $x_{j0k} + x_{j1k} + \ldots + x_{jm-1k}$ constitutes the k-th smallest distance from the user j to a located center. If this k-th smallest distance is denoted by $d_{ik(j)}$, then the expression of $d_{ik(j)}$ by the auxiliary 0-1 variables x_{jsk} is clearly shown in the following Fig. 3.

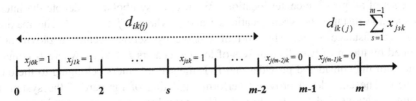

Fig. 3. Expression of the k-th smallest distance from the user j to a located center by the auxiliary 0-1 variables x_{jsk} for $s \in 0 \ldots m - 1$ [13].

Using the above introduced coefficients, ranges and decision variables, we suggest the following model [13].

$$\text{Minimize} \sum_{j \in J} b_j \sum_{s=0}^{m-1} \sum_{k=1}^{r} q_k x_{jsk} \tag{1}$$

$$\text{Subject to} \sum_{i \in I_R} y_i = p \tag{2}$$

$$\sum_{k=1}^{r} x_{jsk} + \sum_{i \in I_R} a_{ij}^s y_i + \sum_{i \in I_F} a_{ij}^s \geq r \qquad \text{for } j \in J, \ s = 0 \ldots m - 1 \tag{3}$$

$$y_i \in \{0, 1\} \qquad \text{for } i \in I_R \tag{4}$$

$$x_{jsk} \in \{0, 1\} \qquad \text{for } j \in J, \ s = 0 \ldots m - 1, \ k = 1 \ldots r \tag{5}$$

The objective function (1) expresses the expected volume of generalized disutility. Constraint (2) ensures that the number of centers belonging to the considered part of the emergency service system under reengineering will be constant. For given pair of user j and a distance value s, the constraint (3) expresses relation between the set of location variables y_i, $i \in I_R$ and the sum of auxiliary variables x_{jsk} over range $1 \ldots r$ of subscript k. If no center is located in the radius s, then the sum of auxiliary variables x_{jsk} equals to r. If exactly $k \leq r$ centers are located in the radius s, then the sum of variables equals to $r - k$ due to minimization process, which presses down the values of the variables x_{jsk}. If the sum of variables x_{jsk} equals to $k < r$, then the variables x_{js1}, ..., x_{jsr-k}, equal to zero and remaining variables equal to one due to the used optimization process and decreasing values of the coefficients q_1, ..., q_r.

Then, the objective function value of the optimal solution of the problem (1)–(5) gives expected total length or time of trips from the service centers to the demand locations necessary for satisfaction of all demands for service. This holds subject to assumptions that the coefficients $q_1 \dots q_r$ correspond to the probability values expressing that the k-th nearest center is the first available (unoccupied) service center and that demand volume b_j is proportional to the number of trips necessary for the demand satisfaction. The model (1)–(5) is much more realistic than the original approach based on the simple weighted p-median problem, which corresponds to the case of $r = 1$. The bigger accuracy of the model (1)–(5) is paid for by higher complexity of the solved problem, which may issue to enormous increase of computational time. We want to find, what limit of accuracy presented by the value of r pays off regarding the increase of computational time. As a solution of the problem (1)–(5) is discrete and the values of probabilities q_k sharply decrease, we think that influence of increasing value of r may appear negligible behind some limiting value r^*.

3 Reengineering Under Auxiliary Constraints

As mentioned in Sect. 1 and in [12, 13], the administrator of the system sets up parameters of rules to prevent a designer of new center deployment from increasing provider's benefit at the expense of the system users. The rules must have a simple form to be easy to evaluated and checked. The first rule limits the total number w of the provider's centers, which locations can be changed. The second rule limits the distance between current and newly suggested location of a service center.

To be able to formulate the rules in a concise way, we derive several auxiliary structures using Fig. 4. We assume that all points 1–11 represent system users and the black points 2, 3, 9 and 11 represent current service center locations.

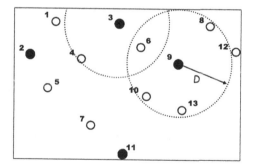

Fig. 4. Simple example of reengineering restrictions [13].

Let $N_t = \{i \in I_R : d_{ti} \leq D\}$ denote the set of all possible center locations, to which the center $t \in I_L$ can be moved subject to limited length of the move. If we consider the example depicted in Fig. 4, we can observe that the center located at the point 9 can be moved to 6, 8, 10 and 13 or stay unchanged. Thus, the set $N_9 = \{6, 8, 9, 10, 13\}$.

Similarly, $N_3 = \{3, 4, 6\}$. Additionally, let $S_i = \{t \in I_L : i \in N_t\}$ denote a set of all centers of the considered provider, which can be moved to $i \in I_R$ subject to the mentioned limitation. Here $S_6 = \{3, 9\}$. Realize that $t \in N_t$ and $i \in S_i$ for $t \in I_L$ and $i \in I_R$ and thus $I_L \subset I_R$.

Now, we introduce series of decision reallocation variables, which model the decisions on moving centers from their original positions to new ones. The variable $u_{ti} \in \{0, 1\}$ for $t \in I_L$ and $i \in N_t$ takes the value of one, if the service center at t is to be moved to i and it takes the value of zero otherwise. Using the above introduced structures and variables we suggest the following model extension.

$$\sum_{i \in I_L} y_i \geq p - w \tag{6}$$

$$\sum_{i \in N_t} u_{ti} = 1 \quad for\ t \in I_L \tag{7}$$

$$\sum_{t \in S_i} u_{ti} \leq y_i \quad for\ i \in I_R \tag{8}$$

$$u_{ti} \in \{0, 1\} \quad for\ t \in I_L,\ i \in N_t \tag{9}$$

Constraint (6) limits the number of changed center locations by the constant w. Constraints (7) allow moving the center from the current location t to at most one other possible location in the radius D. Constraints (8) enable to bring at most one center to a location i subject to condition that the original location of the brought center lies in the radius D. These constraints also assure consistency among the decisions on move and decisions on center location.

Another simpler modelling approach to the formal rules implementation consists in relaxation of the parameter D and associated constraints (7)–(9). Parameter D is used to limit the radius, within which an existing service center can be relocated. The idea introduced in [12] assumes that this limitation may represent too strict constraint. This simplified approach enables us to exclude the variables $u_{ti} \in \{0, 1\}$ from the model. The relaxation consists in the fact that the system reengineering is performed in such a way that there must be at least one center located in the radius D from each existing service center and the center relocation will not limited by any distance. Based on these preliminaries, the constraints (7)–(9) may be replaced by the following expression (10), which guarantees that there will be at least one service center located in radius D from each currently located center. Then the associated simplified radial model takes the form of (1)–(6) and (10).

$$\sum_{i \in N_t} y_i \geq 1 \quad for\ t \in I_L \tag{10}$$

The advantage of this simplified model consists in less number of decision variables and structural constraints. Thus, the problem is expected to be easily solvable due to lower model complexity. Furthermore, this model allows such center relocation that is shown in Fig. 5. Consider a transportation network, in which we have two centers (see the left part of the figure). If the optimal solution of the reengineering model places one of the current centers (the green one) to the red node, all the structural constraints stay met and the other center (the blue one) may be relocated anywhere without additional restrictions. Such solution would not be feasible in the original approach described by the model (1)–(9).

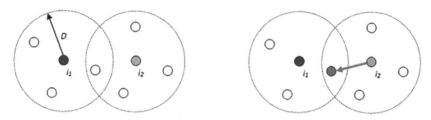

Fig. 5. Reengineering of current emergency system using the simplified model [12]. (Color figure online)

The simplified model (1)–(6), (9) provides bigger variety from the point of possible center location changes in comparison to the original model (1)–(9) and thus it is expected that it could bring better results concerning the optimization criterion. On the other hand, even if this approach may bring better results concerning optimization, the obtained system design can be bad from the point of real system performance. If too many system users are assigned to the same service center, then it is obvious that the service center will be mostly unavailable due to its limited capacity. Therefore, we do not deal with this simplified model in our computational study and we take into account the original formal rules given by parameters w and D.

We want to answer the question about technical solvability of the formulated problem (1)–(9). We ask whether a common commercial integer programming solver based on the branch-and-bound technique is able to find the exact solution of a real-sized problem in acceptable time. Addition of new structural constraints to the original radial model is always questionable from the point of problem solvability. It may directly influence the computational process convergence. Furthermore, we have to realize that even if the administrator's rules are established to defend users' interests, they may represent a reduction of the set of feasible solutions. This phenomenon may lead to a less possible benefit (higher disutility) for the average user. That is why, the dependence of the optimal objective function value on setting of parameters w and D is worth to study.

4 Numerical Experiments

4.1 Benchmarks and Preliminaries

The numerical experiments in this section were performed in the optimization software FICO Xpress 8.3, 64-bit. The experiments were run on a PC equipped with the Intel® Core™ i7 5500U 2.4 GHz processor and 16 GB RAM.

Used benchmarks were derived from real emergency health care system, which was implemented in eight regions of Slovak Republic. For each self-governing region, i.e. Bratislava (BA), Banská Bystrica (BB), Košice (KE), Nitra (NR), Prešov (PO), Trenčín (TN), Trnava (TT) and Žilina (ZA), all cities and villages with corresponding number of inhabitants b_j were taken into account. The coefficients b_j were rounded to hundreds. The set of communities represents both the set J of users' locations and the set I of possible center locations as well. The cardinalities of these sets vary from 87 to 664 locations. The total number of possible service center locations for the individual self-governing region is reported in Table 1 in the column denoted by $|I|$. Each self-governing region emergency sub-system provides its user with emergency service from the given number of service centers currently located at some of the possible locations from I. The number of service centers of the individual region is reported in Table 1 in the column denoted by TNC (the Total Number of located Centers). In all solved instances, we consider that disutility perceived by a system user is represented by response time and this response time is proportional to the network distance, which must be traversed from the servicing center to the user locations. As the generalized disutility according to the model in Sect. 2 is studied in this paper, associated parameters r and q_k for $k = 1 \dots r$ must be established. For these numerical experiments, the value of r was set to 7 and the associated coefficients q_k for $k = 1 \dots r$ were set in percentage in the following way: $q_1 = 77.063$, $q_2 = 16.476$, $q_3 = 4.254$, $q_4 = 1.593$, $q_5 = 0.47$, $q_6 = 0.126$, and $q_7 = 0.018$. These values were obtained from a simulation model of existing emergency medical service system in Slovakia [9].

To enrich the pool of benchmarks, we created ten instances for each self-governing region so that they differ in the list of located service centers operated by the considered provider. The average percentage rate of the provider's centers is reported in Table 1 in the column denoted by "Prov. [%]".

5 Basic Experiments

The basic experiments reported in this sub-section were originally published in [13]. An individual experiment was organized so that the optimal solution of the problem (1)–(5) was obtained first. The solution does not represent reengineering subject to auxiliary rules specified in Sect. 3, but it represents the best possible solution of the emergency system design problem without any restrictions. The average results are summarized in Table 1, where the right part of the table contains the average computational times in seconds across the ten instances solved for each region. The average computational times are reported in the column denoted by "CT [s]". The last column "$ObjF$" contains the average values of the objective function (1).

The results indicate that the reengineering of the emergency service system under generalized disutility for $r = 7$ from the point of service provider does not represent a hard solvable problem. It can be observed that the radial formulation enables to get the optimal solution within 1 min.

Table 1. Average results of numerical experiments for each self-governing region. The value of r was set to 7 [13].

Reg.	$|I|$	TNC	Prov. [%]	CT [s]	$ObjF$
BA	87	14	55.1	0.5	28087.8
BB	515	36	44.9	43.6	47706.5
KE	460	32	46.0	30.4	48490.9
NR	350	27	50.7	10.8	52024.6
PO	664	32	44.3	50.5	61070.2
TN	276	21	52.9	5.1	36800.9
TT	249	18	49.6	6.1	43986.1
ZA	315	29	46.8	6.1	45341.2

In spite of this useful feature, the following portion of experiments was performed to find out, whether a lower value of r will have significant influence on the resulting objective function value. As mentioned in Sect. 2, we assume that the influence of increasing value of r may appear negligible behind some limiting value r^*. To confirm this hypothesis and to find a suitable value of r^*, we have solved all instances with different values of r. For bigger comfort of computation, we expressed the probabilities q_k in percentage, i.e. their sum equals to one hundred. If $r < 7$, then the coefficient q_r was computed according to (11) as a complement of the coefficients q_k for $k = 1 \ldots r-1$ to the value of 100, i.e. the sum of q_k for $k = 1 \ldots r$ must equal 100.

$$q_r = 100 - \sum_{k=1}^{r-1} q_k \tag{11}$$

The dependency of average computational time on the value of r was studied first. We assume that the computational time increases with growing value of r, because the value of r affects the number of variables and the model size as well. Our expectation has been confirmed by the results summarized in Table 2. Each row of the table represents the average results of the ten instances for each region and the columns are used for different setting of parameter r. The last row contains the average values of all instances. The dependency of average computational time on the value of r is also shown in Fig. 6.

Table 2. Average computational time in seconds of the solving process depending on r for each region [13].

Reg.	$r = 1$	$r = 2$	$r = 3$	$r = 4$	$r = 5$	$r = 6$
BA	0.1	0.1	0.2	0.2	0.3	0.3
BB	6.5	8.3	11.5	19.3	27.1	33.7
KE	6.0	7.0	8.9	11.6	15.7	18.3
NR	2.1	2.6	3.6	6.7	6.8	8.4
PO	20.6	22.8	26.1	31.8	38.8	47.0
TN	1.4	1.8	3.2	2.8	3.4	4.3
TT	1.2	1.7	2.3	3.1	4.2	5.2
ZA	1.7	2.0	2.6	3.3	4.4	5.2
AVG	**4.96**	**5.79**	**7.29**	**9.87**	**12.58**	**15.29**

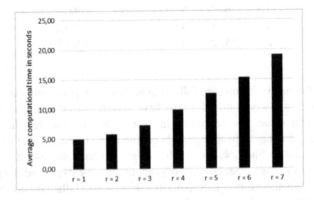

Fig. 6. Dependency of average computational time in seconds on the number r [13].

When we studied the impact of the number r of considered centers on the resulting system design, we have evaluated Hamming distance of the vectors of resulting location variables obtained for various values of the parameter r. The Hamming distance of two 0-1 vectors y and z is defined by the expression (12). The average results are reported in Table 3.

$$HD(y, z) = \sum_{i \in I} (y_i - z_i)^2 \tag{12}$$

The dependency of average Hamming distance from the optimal solution obtained for $r = 7$ on the number r of service providing centers for each system user is also shown in Fig. 7.

The reported results show that the suitable value of r^* is 3. Thus, we proved that three nearest located service centers are enough to be taken into account, when emergency system reengineering under generalized disutility is performed. As shown in Table 3 and Fig. 7, the service center deployment for $r = 3$ differs from the service center deployment obtained for $r = 7$ only in one center on the average. The difference in one service center corresponds to Hamming distance equal to the value of two.

Table 3. Average Hamming distance from the optimal solution obtained for $r = 7$ computed for each region [13].

Reg.	$r = 1$	$r = 2$	$r = 3$	$r = 4$	$r = 5$	$r = 6$
BA	5.2	2.0	0.6	0.4	0.0	0.0
BB	12.6	11.0	3.6	0.6	0.4	0.0
KE	11.8	6.0	3.4	1.4	0.6	0.6
NR	10.2	6.6	2.0	0.4	0.0	0.0
PO	11.8	7.4	4.0	0.0	0.6	0.0
TN	8.0	2.6	0.6	0.2	0.0	0.0
TT	8.4	3.8	0.4	1.4	0.0	0.0
ZA	12.4	4.0	3.2	1.0	0.2	0.0
AVG	**10.05**	**5.43**	**2.23**	**0.68**	**0.23**	**0.08**

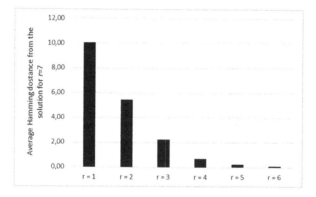

Fig. 7. Dependency of Hamming distance from the optimal solution obtained for $r = 7$ on the number r of service providing centers for each system user [13].

The last characteristics studied in this portion of experiments is the objective function value. For each system design obtained for particular value of $r = 1, …, 6$, the objective function (1) with $r = 7$ and the full set of coefficients q_k was computed. This value was compared to the objective function value obtained for $r = 7$ and the gap between the objective values was evaluated. The gap is defined as a percentage difference of the two objective function values, where the objective function value for $r = 7$ was taken as the base. The average values of gaps of the ten instances computed for each self-governing region are reported in Table 4, which is organized in the same manner as the previous tables. To enable finding a suitable value of $r*$, the gaps lower than 0.1% are marked by grey color.

The dependency of average gap from the optimal solution obtained for $r = 7$ on the number r of service providing centers for each system user is shown also in Fig. 8.

The detailed analysis of presented results shows that usage of the three nearest service providing centers instead of the nearest seven centers leads to very similar results and saves more than one half of computational time. That is why, the next experiments were performed with the setting of $r = 3$.

Table 4. Average gap from the optimal solution obtained for $r = 7$ [13].

Reg.	$r = 1$	$r = 2$	$r = 3$	$r = 4$	$r = 5$	$r = 6$
BA	2.52	0.23	0.02	0.01	0.00	0.00
BB	6.19	0.43	0.07	0.00	0.00	0.00
KE	2.88	0.21	0.11	0.01	0.00	0.00
NR	2.31	0.55	0.06	0.00	0.00	0.00
PO	5.19	0.62	0.04	0.00	0.00	0.00
TN	2.81	0.24	0.04	0.01	0.00	0.00
TT	2.60	0.32	0.01	0.02	0.00	0.00
ZA	4.33	0.69	0.05	0.00	0.00	0.00
AVG	**3.73**	**0.43**	**0.05**	**0.01**	**0.00**	**0.00**

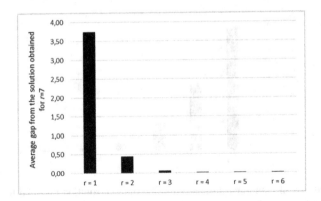

Fig. 8. Dependency of average gap from the optimal solution obtained for $r = 7$ on the number r of service providing centers for each system user [13].

The following table contains the final comparison of current service center deployment to the results of suggested reengineering model, which was configured as follows. Based on the above presented results, the parameter r was set to 1 (simple disutility) and 3 (generalized disutility). In the experiments with the generalized disutility, the associated probability coefficients $q_1 = 77.063$, $q_2 = 16.476$ and $q_3 = 6.461$ were used. The objective function values reported in the table was recomputed for setting $r = 7$ and the original seven probability values. The maximal number w of centers operated by the considered service provider, which are allowed to change their current location, was set to the cardinality of the set I_L, i.e. locations of all considered provider's centers could be changed. The value 15 limited the radius D, in which a center could be relocated. This initial value of D corresponds to the rule applied in the emergency health care system of the Slovak Republic [12]. Table 5 contains the average results of the ten instances solved for each self-governing region. The objective function value of the current service center deployment is reported in the column denoted by "Current $ObjF$". The right part of the table is dedicated to the results of suggested reengineering problems. The abbreviation "$ObjF$" denotes the objective function value of the emergency system design obtained by solving the reengineering model. Finally, the value of Imp was computed to show possible improvement of the objective function value expressed by the generalized disutility, which can be achieved by relocating $w = |I_L|$ service centers. The value of possible improvement was computed as a percentage difference between objective function values of the current service center deployment and the new system design resulting from optimal solution of the problem described by the associated model. The objective function value of current deployment was taken as the base.

Table 5. Comparison of current service center deployment to the results of reengineering model for $r = 1$ (simple disutility) and $r = 3$ (generalized disutility). The reengineering parameters were set at $w = p$ and $D = 15$ [13].

Reg.	Current $ObjF$	$r = 1$		$r = 3$	
		$ObjF$	$Imp.$ [%]	$ObjF$	$Imp.$ [%]
BA	29792	28810	3.30	28335	4.89
BB	52510	51094	2.70	50430	3.96
KE	52786	51894	1.69	50913	3.55
NR	56759	54440	4.09	53472	5.79
PO	67037	65807	1.83	63526	5.24
TN	38625	38091	1.38	37226	3.62
TT	472163	45569	3.48	44734	5.25
ZA	49324	47566	3.56	46634	5.45

The reported results show that the emergency system reengineering can cause a considerable improvement of service accessibility for system users expressed by general disutility. The average values of the improvement (Imp) indicate that the objective function value corresponding to the system design can be reduced up to 6%. The achieved results also confirm the usefulness of suggested reengineering model, because it enables us to obtain better system design from the point of service accessibility. It is obvious from the comparison of the cases with $r = 1$ and $r = 3$ that the usage of generalized disutility ($r = 3$) leads to such solutions, which are approximately by 2% better than those, which can be obtained by usage of simple disutility model ($r = 1$).

5.1   Extended Experiments

This portion of experiments was devoted to observation of impact of the additional constraints (6)–(9) on the solving process. We concentrated on the three parameters, which may play role both in reengineering effectiveness and computational time. This set of experiments was organized so that two of the parameters were fixed and the third one was changed across a given range. The studied parameters were: p – the number of centers supervised by the considered provider, w – the total number of provider's centers, which can be moved, and D – the maximal radius, in which a center can be moved.

In the first part of this portion of experiments, the parameter p was set at the value reported at the beginning of this section, i.e. $p = |I|/2$ and the experiments were performed either with fixed parameter D or parameter w. This portion of experiments was originally published in [13].

First, the maximal radius D was fixed at the value of 15 and the maximal number w of centers allowed to change their locations was set to $p/4$, $p/2$, $3p/4$, and p respectively. Dependency of average computational time in seconds computed for the ten instances of each region is reported in Table 6. Each row of the table represents one region and the columns are devoted to different settings of w.

Table 6. Average computational time in seconds for each region and different values of w. Parameter D was 15 [13].

Reg.	$w = p/4$	$w = p/2$	$w = 3p/4$	$w = p$
BA	0.11	0.11	0.11	0.12
BB	4.36	6.40	5.36	5.32
KE	4.44	5.95	5.77	5.09
NR	2.02	2.20	2.75	2.78
PO	9.78	9.76	9.79	9.83
TN	1.55	1.64	1.68	1.73
TT	1.30	2.30	1.47	1.52
ZA	1.74	1.65	1.63	1.66
AVG	3.16	3.75	3.57	3.50

The reported results show that different settings of w do not significantly affect the computational process, because the value of w limits only the number of possible service center location changes and thus, the number of variables and constraints is independent on w.

The objective function values can be studied in Table 7. Even if parameter r was set to 3 in all solved models, the objective function values were computed for $r = 7$ based on the resulting service center deployment.

Second, we fixed the parameter w to its maximal value p, i.e. all centers operated by the provider could change their current locations. Then, we explored the impact of the parameter D on the solving process complexity.

Table 7. Average objective function values for each region and different values of w. Parameter D was 15. The objective function value was recomputed for $r = 7$ and the whole set of probability coefficients q_k [13].

Reg.	$w = p/4$	$w = p/2$	$w = 3p/4$	$w = p$
BA	28607.0	28334.8	28334.8	28334.8
BB	50676.2	50433.7	50430.4	50430.4
KE	51141.3	50916.9	50913.4	50913.4
NR	53995.8	53482.7	53471.5	53471.5
PO	63791.3	63532.1	63526.0	63526.0
TN	37286.6	37225.5	37225.5	37225.5
TT	45670.3	44915.7	44733.6	44733.6
ZA	47278.1	46673.0	46634.3	46634.3

The average computational times in seconds computed for each self-governing region and given values of D across the range of 5, 10, 15, 20, and 25 are reported in Table 8.

Table 8. Average computational time in seconds for each region and different values of D. Parameter w was set to its maximal value, i.e. $w = p$ [13].

Reg.	$D = 5$	$D = 10$	$D = 15$	$D = 20$	$D = 25$
BA	0.04	0.08	0.12	0.16	0.17
BB	0.90	3.13	5.32	10.54	15.38
KE	1.02	2.61	5.09	7.41	8.81
NR	0.42	1.11	2.78	6.91	5.31
PO	1.91	4.93	9.83	15.85	19.06
TN	0.40	1.05	1.73	2.16	2.96
TT	0.28	0.79	1.52	1.99	2.44
ZA	0.45	0.97	1.66	2.14	2.81
AVG	0.68	1.83	3.50	5.89	7.12

The results reported in Table 8 have confirmed our expectation that the parameter D has a direct impact on the computational process. As it can be observed, the average computational time grows with increasing value of D, i.e. with increasing radius, in which current center can be removed. This phenomenon has a simple explanation. The bigger is the radius for center location change, the higher is the number of its possible new locations. As we can notice, this parameter defines the number of decision variables and it directly affects the model size. Therefore, the solving process takes longer time for higher distance D. Finally, the dependency of objective function value on the parameter D is shown in Table 9.

Table 9. Average objective function values for each region and different values of D. Parameter $w = p$. The objective function value was recomputed for $r = 7$ and the whole set of probability coefficients q_k [13].

Reg.	$D = 5$	$D = 10$	$D = 15$	$D = 20$	$D = 25$
BA	29563.0	28798.3	28334.8	28255.0	28136.1
BB	52115.0	50635.4	50430.4	49429.6	49130.0
KE	52111.9	51398.4	50913.4	50412.6	49959.5
NR	56153.7	54360.2	53471.5	52674.5	52422.9
PO	66115.8	65081.5	63526.0	63070.1	62444.1
TN	37714.0	37320.5	37225.5	37148.4	37009.5
TT	46162.7	45395.9	44733.6	44114.7	44078.5
ZA	48712.7	47763.3	46634.3	46230.4	46115.0

As far as the objective function value expressed by generalized disutility is concerned, the achieved results indicate that the higher value of D is, the better solution can be obtained. As the radius D defines the set of new possible center locations, its setting affects the possibility for obtaining better results. More elements in the set N_t for each $t \in I_L$ causes more candidates for new center locations and bigger possible change of current center deployment, which can be followed by better service accessibility for system users.

To verify obtained results and to confirm observed dependences, another portion of numerical experiments was suggested. This new set of experiments was performed for a new set of benchmarks. The new instances were generated from the same transportation networks as before, but these instances differ in the percentage of service centers operated by the considered provider. While in the preliminary experiments, the considered provider owned approximately half of all located centers, in these new benchmarks the ratio is 25 and 75% respectively. The reported results represent the average values of 10 problem instances. Since the self-governing region of Bratislava (BA) is too small for such a study, it was excluded from this portion of experiments.

The main goal of these experiments was to study the impact of formal parameters w and D on the computational time and the resulting system design quality measured by the value of generalized disutility. The obtained results are summarized in the following eight tables. Tables 10, 11, 12 and 13 contain the results aimed at studying the impact of individual parameters w and D on the average computational time. The structure of the tables is the same as used in Tables 6 and 8.

Finally, the last set of tables reports the studied impact of individual parameters w and D on the optimization criterion, which consists in generalized disutility perceived by an average system user. The obtained results are summarized in Table 14, Table 15, Table 16 and Table 17, which follow the structure of Table 7 and Table 9 respectively.

Table 10. Average computational time in seconds for each region and different values of w. Parameter D was 15. The considered provider operated 25% of all located service centers.

Reg.	$w = p/4$	$w = p/2$	$w = 3p/4$	$w = p$
BB	2.2	2.2	2.1	2.2
KE	2.1	3.0	3.1	2.7
NR	1.1	1.3	1.3	1.3
PO	4.9	5.0	5.0	5.2
TN	0.7	0.7	0.7	0.8
TT	0.7	0.7	0.8	0.8
ZA	0.9	0.9	0.9	0.9

Table 11. Average computational time in seconds for each region and different values of w. Parameter D was 15. The considered provider operated 75% of all located service centers.

Reg.	$w = p/4$	$w = p/2$	$w = 3p/4$	$w = p$
BB	22.7	29.1	13.2	13.0
KE	10.0	28.2	33.1	27.7
NR	7.8	9.4	9.2	10.9
PO	36.6	36.4	37.0	36.9
TN	3.3	3.5	3.6	3.6
TT	4.8	4.0	4.1	4.2
ZA	6.7	4.5	4.6	4.7

Table 12. Average computational time in seconds for each region and different values of D. Parameter $w = p$. The considered provider operated 25% of all located service centers.

Reg.	$D = 5$	$D = 10$	$D = 15$	$D = 20$	$D = 25$
BB	0.7	1.4	2.2	3.5	4.7
KE	0.7	1.3	2.7	4.4	6.0
NR	0.3	0.6	1.3	1.8	2.0
PO	1.4	2.8	5.2	9.0	12.1
TN	0.3	0.5	0.8	1.0	1.3
TT	0.2	0.4	0.8	1.1	1.5
ZA	0.3	0.6	0.9	1.2	1.5

Table 13. Average computational time in seconds for each region and different values of D. Parameter $w = p$. The considered provider operated 75% of all located service centers.

Reg.	$D = 5$	$D = 10$	$D = 15$	$D = 20$	$D = 25$
BB	1.4	6.1	13.0	48.4	48.8
KE	1.6	6.8	27.7	13.1	22.8
NR	0.7	10.5	10.9	26.4	28.8
PO	4.3	19.7	36.9	85.5	58.1
TN	0.6	2.1	3.6	6.2	6.8
TT	0.4	2.0	4.2	5.2	5.3
ZA	0.8	2.7	4.7	6.5	7.6

Table 14. Average objective function values for each region and different values of w. Parameter D was 15. The objective function value was recomputed for $r = 7$ and the whole set of probability coefficients q_k. The considered provider operated 25% of all located service centers.

Reg.	$w = p/4$	$w = p/2$	$w = 3p/4$	$w = p$
BB	51867.3	51774.2	51774.1	51774.1
KE	52321.0	52168.6	52156.7	52142.6
NR	55479.0	55290.3	55291.7	55291.7
PO	65226.3	64901.0	64882.2	64882.2
TN	38140.5	38089.5	38089.5	38089.5
TT	46322.2	46009.7	45934.3	45920.1
ZA	48676.3	48448.5	48412.9	48412.9

Table 15. Average objective function values for each region and different values of w. Parameter D was 15. The objective function value was recomputed for $r = 7$ and the whole set of probability coefficients q_k. The considered provider operated 75% of all located service centers.

Reg.	$w = p/4$	$w = p/2$	$w = 3p/4$	$w = p$
BB	49801.2	49513.4	49512.0	49512.0
KE	50148.0	49918.1	49915.2	49915.2
NR	52756.2	51979.4	51969.4	51969.4
PO	62266.2	61691.3	61691.3	61691.3
TN	36987.9	36932.8	36932.8	36932.8
TT	44978.0	44017.9	44007.3	44007.3
ZA	45697.8	44883.3	44873.4	44873.4

Table 16. Average objective function values for each region and different values of D. Parameter $w = p$. The objective function value was recomputed for $r = 7$ and the whole set of probability coefficients q_k. The considered provider operated 25% of all located service centers.

Reg.	$D = 5$	$D = 10$	$D = 15$	$D = 20$	$D = 25$
BB	52324.8	51838.4	51774.1	51374.3	51106.5
KE	52482.0	52346.0	52142.6	51892.2	51401.5
NR	56551.4	55592.9	55291.7	54573.7	54199.4
PO	66364.2	65718.3	64882.2	64501.3	63863.7
TN	38208.0	38089.5	38089.5	38087.1	38087.1
TT	46799.8	46175.2	45920.1	45570.9	45570.9
ZA	49007.2	48681.2	48412.9	48125.2	48055.8

Table 17. Average objective function values for each region and different values of D. Parameter $w = p$. The objective function value was recomputed for $r = 7$ and the whole set of probability coefficients q_k. The considered provider operated 75% of all located service centers.

Reg.	$D = 5$	$D = 10$	$D = 15$	$D = 20$	$D = 25$
BB	51954.4	49950.7	49512.0	48343.0	47849.3
KE	51716.4	50681.5	49915.2	48744.4	48558.1
NR	55850.9	53368.5	51969.4	51013.8	50878.2
PO	65139.8	63550.4	61691.3	60774.1	59844.4
TN	37409.3	36989.8	36932.8	36836.8	36629.7
TT	45987.3	45064.7	44007.3	43518.4	43369.4
ZA	48277.4	46159.8	44873.4	44563.7	44508.3

5.2   Mutual Relation of the Formal Rules

All the experiments presented above were aimed primarily at studying the model solvability and the sensitivity of the associated computational process on different model parameters. Besides some interesting findings and suitable settings of parameters, we focused also on mutual combinations of the administrator's rules and their impact on the computational time and the quality of resulting solution given by generalized disutility. The considered rules are that at most given number of center locations can be changed and each center location can be moved only in a given radius from its original position. For these experiments, the self-governing region of Žilina was used. Here, it was assumed that the considered provider operates half of all service centers.

An individual experiment was organized so that the reengineering was performed using the model (1)–(9) for different values of parameters w and D. The parameter w expresses the number of service centers, which can change their current location. Parameter D limits the radius, in which the service center can be relocated. This way, 20 problems for all combinations of mentioned parameters were solved for each problem instance. The results obtained for the individual self-governing regions are presented in Table 18 and Table 19 respectively. The first table contains the average computational time in seconds. It must be noted that ten different instances were generated randomly for each self-governing region as described in previous sections. These instances differ in the list of located service centers operated by the considered provider. The parameter w was set to 25, 50, 75 and 100% of the total number of centers operated by the considered provider. The parameter D took the value 5, 10, 15, 20 and 25.

The results reported in Tables 18 and 19 have proved our expectations and confirmed previously observed trends. This new portion of experiments was focused on studying the efficiency of the administrator's rules imposed on provider's changes. It was found that the parameters w and D may directly influence the resulting system design, because they affect possible changes in current service center deployment. As far as computational time is concerned, parameter w does not have significant impact. It is used as the right side of the constraint (6) and thus, its value does not change the model size. On the other hand, parameter D considerably affects the solving process. It must be noted that the radius D defines the set of all new possible locations of a service center and thus, it affects the number of decision variables and structural constraints as well. More elements in the set N_t for each $t \in I_L$ mean more candidates for new center locations and higher computation time.

Table 18. Average computational times in seconds for different settings of parameters w and D in the self-governing region of Žilina (ZA), in which the considered provider operates approximately half of all located service centers.

	$D = 5$	$D = 10$	$D = 15$	$D = 20$	$D = 25$
$w = p/4$	0.5	1.0	1.8	2.7	3.9
$w = p/2$	0.5	1.0	1.7	2.3	2.5
$w = 3p/4$	0.5	1.0	1.7	2.2	3.1
$w = p$	0.5	1.0	1.8	2.3	2.9

Table 19. Average objective function values for different settings of parameters w and D in the self-governing region of Žilina (ZA), in which the considered provider operates approximately half of all located service centers. The objective function value was recomputed for $r = 7$ and the whole set of probability coefficients q_k.

	$D = 5$	$D = 10$	$D = 15$	$D = 20$	$D = 25$
$w = p/4$	48761.3	48043.0	47278.1	46984.8	46826.5
$w = p/2$	48712.7	47773.1	46673.0	46273.6	46159.3
$w = 3p/4$	48712.7	47763.3	46634.3	46230.4	46115.0
$w = p$	48712.7	47763.3	46634.3	46230.4	46115.0

6 Conclusions

This paper deals with emergency medical system reengineering under consideration of generalized disutility, which follows the idea that the associated service can be provided from given number of the nearest located centers. Application of the generalized disutility makes the model more realistic by taking into account possible temporarily unavailability of service centers. In our computational study we have found, that three nearest located centers are enough to be considered in the objective function value, because the accuracy of the result is satisfactory.

The next part of experiments was aimed at exploration of impact of additional constraints imposing new restrictions on service center location changes. The additional constraints regulate extent of the permitted reengineering and the reported results give deeper insight into their influence upon computational time of the solving process and quality of the resulting service system design. Based on the results and obtained experience, we can conclude that we have constructed a very useful tool for emergency medical system reengineering under generalized disutility performed by the system administrator with service centers of a considered provider. Designed and investigated model is easy to be implemented and solved in common optimization environment equipped with the branch-and-bound method or other technique destined for integer programming problems.

Future research in this field may be aimed at such system reengineering, which takes into account uncertainty following from randomly occurring failures in the underlying transportation network.

Acknowledgment. This work was supported by the research grants VEGA 1/0342/18 "Optimal dimensioning of service systems", VEGA 1/0463/16 "Economically efficient charging infrastructure deployment for electric vehicles in smart cities and communities", and APVV-15-0179 "Reliability of emergency systems on infrastructure with uncertain functionality of critical elements".

References

1. Avella, P., Sassano, A., Vasil'ev, I.: Computational study of large scale p-median problems. Math. Program. **109**(1), 89–114 (2007)
2. Current, J., Daskin, M., Schilling, D.: Discrete network location models. In: Drezner, Z., et al. (eds.) Facility Location. Applications and Theory, pp. 81–118. Springer, Berlin (2002)
3. Doerner, K.F., et al.: Heuristic solution of an extended double-coverage ambulance location problem for Austria. CEJOR **13**(4), 325–340 (2005)
4. Elloumi, S., Labbé, M., Pochet, Y.: A new formulation and resolution method for the p-center problem. INFORMS J. Comput. **16**, 84–94 (2004)
5. García, S., Labbé, M., Marín, A.: Solving large p-median problems with a radius formulation. INFORMS J. Comput. **23**(4), 546–556 (2011)
6. Ingolfsson, A., Budge, S., Erkut, E.: Optimal ambulance location with random delays and travel times. Health Care Manage. Sci. **11**(3), 262–274 (2008)
7. Janáček, J.: Approximate covering models of location problems. In Lecture Notes in Management Science: Proceedings of the 1st International Conference ICAOR, Yerevan, pp. 53–61 (2008)
8. Janáček, J., Kvet, M.: Public service system design with disutility relevance estimation. In Mathematical Methods in Economics, Jihlava, pp. 332–337 (2013)
9. Jankovič P.: Calculating reduction coefficients for optimization of emergency service system using microscopic simulation model. In: In 17th International Symposium on Computational Intelligence and Informatics, Budapest, pp. 163–167 (2016)
10. Jánošíková, Ľ.: Emergency medical service planning. Commun. Sci. Lett. Univ. Žilina **9**(2), 64–68 (2007)
11. Kvet, M.: Computational study of radial approach to public service system design with generalized utility. In: Proceedings of the 10th International Conference on Digital Technologies, pp. 198–208. IEEE (2014)
12. Kvet, M., Janáček, J.: Radiálny prístup na zlepšenie existujúceho záchranného systému. In: Optimalizační úlohy v dopravních a logistických systémech a SW podpora rozhodování v inteligentních dopravních systémech, Praha, pp. 11–25 (2016)
13. Kvet, M., Janáček, J.: Reengineering of the emergency service system under generalized disutility. In: 7th International Conference on Operations Research and Enterprise Systems, ICORES 2018, Madeira, pp. 85–93 (2018)
14. Snyder, L.V., Daskin, M.S.: Reliability models for facility location; the expected failure cost case. Transp. Sci. **39**(3), 400–416 (2005)

Dealing with Scheduling Fairness in Local Search: Lessons Learned from Case Studies

Christophe Ponsard[(✉)] and Renaud De Landtsheer

CETIC Research Centre, Charleroi, Belgium
{christophe.ponsard,renaud.delandtsheer}@cetic.be

Abstract. Many systems undergoing an optimisation process also involve users which might be directly or indirectly impacted in different ways. Fairly spreading this positive or negative impact is required in specific contexts like critical healthcare or due to work regulation constraints. It can also be explicitly requested by users. This papers considers case studies from three different domains involving fairness: night shift planning, clinical pathways and a shared shuttle system. Each case is analysed to understand how fairness requirements were captured, how the solution was designed and implemented. It also analyse how fairness was perceived by the user using the deployed system. We also draw some lessons learned and recommendations which are discussed in the light of similar work reported in other domains.

1 Introduction

Fairness is a concept intuitively easy to understand because as human we have a natural tend to compare with our pairs. We have experienced a lot "fair" or "unfair" behaviours from early age (e.g. candy distribution at school) and are regularly confronted to it in our personal lives (e.g. queuing to enter pass some access control) and at work (e.g. time schedules for teachers, night shifts). It is however difficult to give a precise definition because fairness is a generic name for a multitude of concepts including socio-political ones involving impartiality, justice and equity [12,29]. Fairness goes beyond the pure behaviour design of a system but also involves people perception about its inner "rules" and how well they are enforced. This means the need to give clear and transparent explanations. For example a survey about an organ transplantation system reported it was evaluated as fair by only 29% of the respondents while 29% were unsure and 41% believed it was not fair. However when questioned about their understanding of the system about one third recognised they had a partial understanding and another third not at all [4].

The scope of this paper is the optimisation of large and/or complex systems, typically involving resource allocation to people, hence quite systematically requiring to manage fairness issues, generally in a specific context (business domain, existing work regulation or even company culture). Typical examples

© Springer Nature Switzerland AG 2019
G. H. Parlier et al. (Eds.): ICORES 2018, CCIS 966, pp. 220–243, 2019.
https://doi.org/10.1007/978-3-030-16035-7_12

are people scheduling problems (work shifts, patient planning, etc) or vehicle routing problems (pick-up and delivery, dispatching of technical support, etc).

From the technical point of view, the optimisation of such systems is often dealt with using local search (LS) techniques because of their better ability to scale than other techniques like constraint programming (CP) especially when using efficient incremental computation of invariants using constraint-based local search (CBLS) [39]. Although LS do not provide guarantee of optimality, specific meta-heuristics in the above domains generally result in solutions which are only a few percents away from the optimal with also a trade-off between search time and solution quality [38]. We will focus on such search techniques.

This paper does not propose a "one fits it all approach" to this complex problem but takes a practical approach by analysing three case studies based on real-world system. Those cases are from totally different domains but were quite challenging w.r.t. fairness issues, more precisely:

- NiceWatch is a system to organise night watches among a pool of doctors [22]. Complex rule regulate night work. Moreover all night watches do not have the same value for the doctors, e.g. resting period that follows the watch can extend a week-end or result in an extra day off [17].
- PIPAS is an optimisation engine for managing clinical pathways supporting chemotherapy cycles delivered in day clinic [35]. The fairness has here a crucial medical meaning as the system must guarantee the timed treatment delivery to a large pool of patients [34].
- SAMOBI is an optimisation engine for a sustainable shared shuttle service [25,33]. Fairness is present at different levels: for the driver schedules but also for user which will be priced against the level of sharing or delay/detour they are ready to accept.

This paper focuses on the fairness viewpoint while our previous publication [17,33,34] were concerned by reporting on a specific solution dealing with the whole set of requirements, with fairness being one of the common requirements. In this work, we analyse those previous case to understand how fairness was identified, designed, implemented and most importantly how successful the proposed approach was from the user perception point of view. Taking a step back, we present our lessons learned and recommendations for others having to cope with fairness issues in optimising their systems. In addition, we also take into account techniques and experience reported in other domains, including specific fairness indicators such as max-min (initially defined for fair bandwidth distribution in computer networks) [3]), the Jain indicator reaching its maximal value (1) when fair allocation is achieved [16] and the Global Gini Indicator (GGI) from widely used in economics [41].

The paper is structured as follows. Sections 2, 3 and 4 present each case study for the three domains described here above. Those sections share a similar structure starting with a context presentation, then expressing the fairness constraint, designing a solution and finally analysing the resulting running system, with a specific focus on the user perception. Then Sect. 5 formulates some lessons learned and recommendations. Section 6 discusses them in the light of related work from other domains. Finally Sect. 7 concludes and highlight some future work to further improve the management of fairness constraints.

2   Case Study 1 - Planning of Doctor Night Shifts

2.1   Context

In hospitals, coordination is of prime importance to ensure that necessary workforce is available at the required time and with the required competences. This is because an hospital has to cope with possibly large flows of patients that cannot be interrupted, and medical care requires a set of specific competencies. Medical workforce is needed at different roles throughout the week. These include regular working role, on-site night shifts, and on-call duty. Roles planning must obey a given set of rules about resting time and cope with peoples' unavailabilities. Fairness in the way people are scheduled is also required to maintain a good working atmosphere.

Setting up such planning is time consuming and can be the source of personal frustrations since planning composition, when is performed by a human colleague, can always be perceived as not objective and thus unfair. Fairness in duty rostering is emphasised in guidelines of many countries [27, 28].

2.2   Problem Statement and Fairness Requirements

We detail here the considered rostering problem, with a focus on fairness aspects that can cause doctor frustration.

A *planning* ranges over a given period of time, and defines, for each day of this period of time and each role, who among the set of available doctors will fill in this role. *Roles* can be regular workday at a given place (anaesthetist in operating theatre no 2, emergencies), on-site night duty, or home based on-call duty.

With respect to rostering, doctors are characterized by **a set of legal and organizational attributes** such as contractual availability, qualification, maximal number of duty per month, degree of seniority and assignable duty roles. Besides, there is also a set of personal constraints to be considered, including personal days off that are considered as strong constraints as soon as they are granted, and personal preferences regarding duty roles. These can be positive or negative preferences.

Planning must also Obey a Set of Legal Rules w.r.t. Resting Times:

- a duty role at the hospital lasts for 24 or 25 h. If happening during the week, it starts during the normal working hour starting at 8 am and lasts till 9 am the next day. If it occurs on a Saturday, Sunday or legal holiday, it starts at 9 am and ends at 9 am the next day. A duty role happening between and including Sunday to Thursday is automatically followed by a day off. For a duty role occurring on a Friday, the doctor gets half a day off that can (s)he can place anywhere. For a duty role happening on a Saturday, the next Monday is a day off. Legal holidays are treated as Sundays, and days before legal holidays as Fridays.

- A doctor can only fulfil a single role at a time.
- At any time, among all roles requiring a given qualification, one of them at least must be occupied by a senior doctor. For instance, emergencies and anaesthetist roles.
- A doctor cannot be working or on any form of duty when on holiday or when off its contractual working days in case of part time contract.
- There must be at least five days between two consecutive duties of the same doctor.
- In any period of four weeks, a doctor can have at most one duty occurring during the weekend.
- In case of a part time worker, a resting time following a duty cannot happen when the doctor is not working according to the contract.
- A resting time cannot occur during holiday.

A consequence of the compensation system for duty roles is that some of them are more attractive than others. Duty roles happening on a Thursday are the most attractive, since the doctor gets an extended weekend of three consecutive days. The least attractive roles are the ones happening during the weekend, since the compensation is smaller and they are not covering regular working hours, so that the doctor loses half a day off at the end. Friday and other week days duties are in between. This attractiveness is the main cause of frustration with duty roles assignment. **Planning composition should therefore be fair among doctors about this attractiveness of duties**. Inside a large doctor's team, it's indeed impossible to set up a monthly planning that takes into account all the loads or attractiveness of this peculiar month. One must thus figure out a way to spread out the workload among individuals and over longer time frames.

To summarize and prioritize, a planning must comply with the following elements:

1. first, it must comply with the strong constraints here above;
2. then, the personal preferences must be considered;
3. finally, the attractiveness of the planning must be evenly spread across all doctors.

Some inequities can be tolerated temporarily, but they must be compensated the next month. As discussed in the introduction, the planning must ensure some fairness between doctors, and must propose some mechanism to ensure that doctor have a good perception that the algorithm was fair, even though the planning might trigger some personal frustration.

There is also a requirement that the algorithm must be deterministic. This ensures that the person in charge of triggering the runs of the algorithm does not have the possibility to trigger the algorithm on demand to select a solution that better fits some non-expressed desires. As a consequence, all random functions used in our algorithm, notably to break ties, rely on a deterministic pseudo-random generator.

A last non-functional requirement is the efficiency of the planning engine: it must be able to generate a complete schedule for a single month within a few seconds.

2.3   Designing for Fairness

It is difficult to ensure a good comprehensibility of the algorithm and its execution using existing scheduling engines relying on state-of-the-art algorithms. While those are efficient and can deal with fairness, they are also quite complex and hard to understand for non computer scientists, and they are not designed to provide a traceability of the resulting solution. In order to deal with this issue, **a key decision in this case was to implement a dedicated search engine not relying on any framework** such as Gecode, OR-tools, OscaR, LocalSolver or any others [2,13,30,31].

Attractiveness of the planning is a key element of doctor satisfaction, so that it must be quantified in order to reason upon it. The approach is to define a score of discomfort for each doctor on a given planning. All duties get a score of discomfort; the less attractive, the higher is this score. The discomfort of a doctor for a given planning is the sum of the discomfort of all duties (s)he is assigned to in this planning.

With this mechanism of discomfort score, we can model attractiveness, and compensate inequities from one month to another one, by accumulating the score of discomfort across months.

To find a solution, a greedy approach is used. Because it might fail to fill in a role, we therefore introduced the notion for a role of being unassigned. The key points are summarised here and detailed in the rest of this section.

- The main loop is a simple loop that allocates doctors to roles. It iterates onto unassigned roles, and assigns them to a doctor.
- The role is selected to be among the unassigned ones, as the one that has the fewest possible doctors, in view of the strong constraints and with an ordering based on the expressed preferences (positive or negative). In case of equality, a random role is selected.
- Relaxation is used in case the system reaches a step with no possible assignment. In this case a role that was assigned to a doctor is unassigned to generate the necessary degree of freedom.
- Diversification and cycle detection are also used to avoid the system iterating over the same set of partial allocations, leading to a dead end.

Listing 1.1. Greedy Algorithm for Role Assignment.

```
while ( roleToAsign is not empty ) {
  val currentRole =
    select role in roleToAssign
          minimizing degreeOfFreedom ( role )
  val doctorToAssign =
    select doctor in admissibleDoctor ( currentRole )
          maximizing affinity ( doctor , currentRole )
  assign ( currentRole , doctorToAssign )
  update degree of freedom and admissibleDoctor
}
```

Metaheuristic. The algorithm itself is a greedy approach, with possibility to undo some of the greedy decisions. At each iteration a role is picked up, and a doctor is selected for this role. The algorithm has the possibility to relax an assignment in case of no doctor can be selected for the considered role. Finally, a tabu component is added to prevent relaxing assignments too quickly, and help escape local impossibilities. Basically, the greedy search iterates on roles is a well-chosen order, and assign the current role to a doctor. It is summarised in Listing 1.1.

The iteration on roles is based on the degree of freedom of the role. Roles with the smallest degree of freedom are assigned first. The degree of freedom of a role is the number of doctors that can be assigned to this role, given the strong constraints and the existing assignments. It is updated every time an assignment is performed, or relaxed. A role assignment may impact the degree of freedom of another role because some constraints impose a minimal delay between shifts, notably through a resting period.

Ties for roles as well as doctor selection are broken based on a deterministic pseudo-random selection. We deliberately use a deterministic generator because we want several runs of the algorithm to produce the same output, i.e. to avoid the operator to run the tool several time until some untold constraint is met.

The affinity between a doctor and a role is a weighted sum involving the following elements:

- the preference (positive or negative) between the role and the doctor.
- the attractiveness of the considered role, based on weighting along the features of the role.
- a cumulated satisfaction score of the considered doctor that sums up the attractiveness of the past and already assigned role to this doctor.

Two more technical aspects must be considered to ensure the algorithm can reach a solution:

- **Relaxing in Case of Impossibility.** In case the algorithm reaches a point where the current role cannot be fulfilled by any doctor, one or more assignments are relaxed to provide the necessary freedom to the considered role. The doctor is selected such that he can be assigned to the role after some other assignments are relaxed. This excludes all doctors that are not available this day, based on their contract, for instance. The selected doctor also minimizes the number of assignments that are to be relaxed. Ties are broken based on a deterministic pseudo-random selection, again to ensure that the algorithm is deterministic.

– **Dealing with Allocation and Deallocation Cycles.** The relaxation performed in case of a role cannot be assigned can lead to the algorithm oscillating in a closed loop: a role "a" cannot be assigned, so an assignment involving role "b" is relaxed. In turn to assign role "b" the algorithm can relax role "a", etc. To prevent this, each assignment is added to a "tabu list". This tabu list is determined by setting a number of iterations during which the assigned role cannot be relaxed by the relaxation procedure.

2.4 Validating Fairness

The proposed algorithm was implemented for MedErgo, a Belgian company provides a web-based software application for doctor planning, called *NiceWatch* [22]. The system is composed of a global database containing the actual planning, and a web-based user interface in which doctors can post their own constraints, query their planning, interact with other doctors to barter duty roles, etc. A specific interface is also available to the coordinator to visualize all individual constraints, and set up the planning. This interface is depicted in Fig. 1.

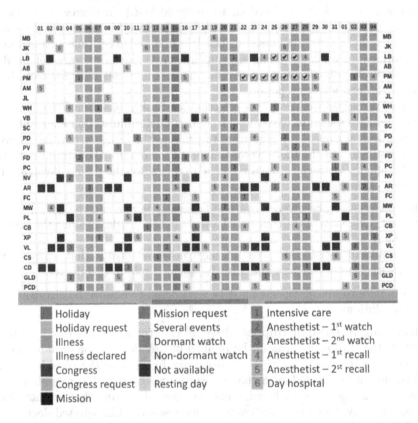

Fig. 1. Planning interface with possible constraints and allocations types [17].

The tool was deployed in production in August 2016 in different Belgian hospitals, as part of the *NiceWatch* web-based platform [22]. Although our solver is not has been developed using standard data structures, the overall performance is quite good as scheduled can be produced within a few seconds and allow the user to wait for the result.

Shifts are typically scheduled each month based on the available doctor staff. The validation example presented here is composed of about 50 doctors which needs to ensure 5 simultaneous watch roles, some at every day, some only on week days. Table 1 shows the staff request for a typical month of 31 days.

Table 1. Typical role request for different watches [17].

	Watch role	# days
1	Intensive care	31
2	Anesthetist – 1st watch	31
3	Anesthetist – 2nd watch	31
4	Anesthetist – 1st watch (recall)	31
5	Anesthetist – 2nd watch (recall)	22
6	Day hospital	22

Figure 1 shows the planning interface displaying both the constraints (coloured square without number) and the proposed allocation (coloured square with number). The legend details the full set of possible constraints and allocations. During the planning process a full allocation trace is generated and available for checking the allocation process. A typical trace is displayed in Listing 1.2 for the first two slots of some allocated day.

Listing 1.2. Justification Trace.

```
4/0/4  Pref=1  P=6.53  Av=30  DDC|  Unavailable  (off)
1/0/1  Pref=1  P=5.90  Av=27  MAF
4/0/4  Pref=1  P=5.74  Av=31  MH
4/0/5  Pref=1  P=8.03  Av=31  DL
4/0/4  Pref=1  P=8.15  Av=31  KS
...
Day X - Slot 1 - Allocated to: MAF

4/0/4  Pref=1  P=8.72  Av=30  CC  |  Unavailable  (off)
4/0/4  Pref=1  P=9.11  Av=31  KB  |  minDistanceKO
1/0/2  Pref=1  P=3.94  Av=21  RCA|  minUnfavDistanceKO
1/0/2  Pref=1  P=4.15  Av=27  SC  |  minDistanceKO
1/0/2  Pref=1  P=4.44  Av=18  WM
1/0/2  Pref=1  P=4.52  Av=22  VN
2/0/3  Pref=1  P=4.86  Av=31  VRP
...
Day X - Slot 2 - Allocated to: WM
```

For each slot, a list reviewing possible doctor allocations can be compiled. The list starts with doctors that cannot be allocated with a justification code whose explanation is detailed in Table 2. It is followed by a prioritised list of doctors using the ranking procedure described in the previous section. According to this clear ranking, the first available doctor is the allocated one and the reason can be easily traced using the justification table.

Table 2. List of Justification in Allocation Traces [17].

Justification	Description
Unavailable	Date is within the strong constraints of the doctor (according to work contract and vacations)
Unwanted	Date is within the doctor's wishes not to be on duty
MaxFrequency Reached	Maximum quota is reached for doctor's wanted watches
MaxFrequency ReachedOutside	Maximum quota is reached for doctor's unwanted watches
RecoveryRule Broken	This day is already assigned or is a recovery day
MinDistanceKO	The minimal delay between two wanted watches cannot be respected
MinDistance UnwantedKO	The minimal delay between two unwanted watches cannot be respected
MinUnfavourable DistanceKO	The minimal delay between two unfavourable dormant watches cannot be respected
MinUnfavourable DistanceOutside KO	The minimal delay between two unfavourable dormant watches cannot be respected
NoSenior	No senior doctor would have been assigned to a set of paired roles
BlackListed	This doctor cannot be assigned here because it will lead to an impossibility to complete the schedule later

3   Case Study 2 - Clinical Pathways for Oncology

3.1   Context

In Western countries, the progress in medical care and the ageing of the population is putting more pressure on hospitals which have to face a growing number of patients. They also need to manage medical procedures of growing complexity and often in a multidisciplinary context. Difficulties to address those challenges can decrease the quality of care received by patients or impact the fairness treatment. A survey of 30 pathologies ranging from osteoarthritis to breast cancer, observed that, on average, only half of the patients received the recommended medical care [21].

To gain better control on care quality, a level of standardisation was proposed through clinical (or care) pathways. A clinical pathway is defined as a multidisciplinary specification of the treatment process required by a group of patients presenting the same medical condition with a predictable clinical course [7]. It describes concrete treatment activities for patients having identical diagnoses or receiving the same therapy.

The goals followed by care pathways are to reach an quality assurance level, keep delays under control but also to reduced operation costs. Because they are strongly process orientated, clinical pathways also provide a global dashboard on the patient journey which overcome the limitation of a collection of specialisation oriented viewpoints gathered through medical records [9].

Clinical pathways have been successfully used for many therapies, such as arthroplasty [40] and breast cancer [9]. In an al oncology context, they require a precise description of the therapeutic workflow and all related activities. Figure 2 depicts a typical workflow for a chemotherapy. It is composed of a sequence of drugs deliveries or cures, usually delivered in day hospital. Each cure is followed by a resting period at home for a duration of some days to some weeks. This resting period is required because a chemotherapy has adverse effect on the whole body with secondary effects like fatigue, pain, mouth and throat sores. If the ideal treatment protocol is followed, the number of cancerous cells should decrease until reaching a where there are no traces of them in the body.

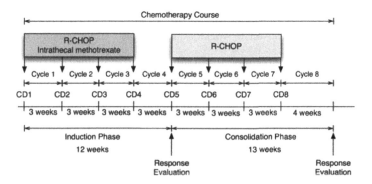

Fig. 2. A typical chemotherapy workflow [35].

If for some reason, chemotherapy cures do not closely follow the intended periodicity or if doses are significantly reduced, the treatment efficiency may be suboptimal. In such conditions, cancerous cells may multiply again, which can result in a cancer relapse.

As a consequence, it is very important to make sure that the care protocol is enforced to a very good level. In order to measure the quality of chemotherapeutic cares, a quantifiable indicator called the "Relative Dose Intensity" (RDI) [19] was defined. It measures both the level of compliance to the required dose and the timing of the delivery, on a scale ranging from 0% (no treatment) to 100% (total conformance).

$$RDI = \frac{\text{planned\_dose}}{\text{delivered\_dose}} x \frac{\text{real\_duration}}{\text{planned\_duration}}$$

To emphasise the importance of adhering to the process, medical survyes have have reported, for a number of cancers, a strong correlation between this RDI and the relapse-free survival time. For breast cancer, a key threshold is 85% [32].

Care pathways are part of a more general evolution toward process-oriented health information systems [15]. To support this evolution, it is necessary to rely on a dedicated scheduling of these workflows because scheduling a large pool of patients in an hospital with limited resources raises a lot of trade-off concerns which are beyond the reach of a human dispatcher [20]. Of course, such concerns should not impact the quality of care of individual patient: the planning should ensure fairness.

3.2   Problem Statement and Fairness Requirements

The problem considered here is the continuous schedule optimisation of an evolving set of patients engaged in a specific chemotherapy process as described in the previous section. The goals are to:

- maintain the best quality of care measured in term of RDI indicator
- meet the resources constraints: available treatment rooms and nurses.
- respect service opening days (weekends, holidays) and hours.
- take into account strong unavailabilities of patients, when known.
- when possible, distribute the workload evenly over time to avoid work peaks.

A clinical pathway is composed of many events that must be managed:

- patients entering and leaving the pathway
- delivery event with possible deviation that can impact the care quality such as partial delivery, advanced/delayed/cancelled delivery, no show...
- medical staff (and possibly rooms) availability

These events are communicated by different actors to the system (e.g. nurses monitoring the drugs delivery, doctors checking the patient condition, administrative staff registering the arrival or non-attendance of a patient). When entering his chemotherapy pathway, a patient is typically given an indicative optimal schedule based on what is known at that time and a confirmation of the first appointment.

To maintain optimality, the occurrence of pathway related events will trigger a re-scheduling. Consequently, the considered scheduling is an *on-line problem* which should meet the following additional constraints:

- the recorded past is of course irreversible: this makes any deviation to the ideal care delivery schedule hard to reverse.
- confirmed appointments for other patients should preferably not be changed because it requires administrative work and can induce a cascading effect.

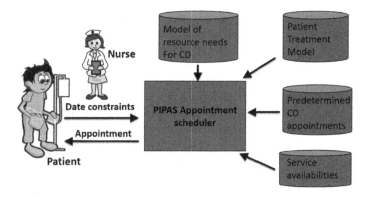

Fig. 3. Problem context diagram [34].

A key actor in charge of activity re-planning is the administrative nurse. He/She is frequently in contact with the patient and acts as a relay between the patient and the system, e.g. to identify and and confirming future delivery dates. A contextual diagram of the information flow between the patient and the pathway management system is depicted in Fig. 3. The required information for the scheduling requires to capture all the required resources needs for the Chemotherapy Deliveries (CD) (i.e. room time, staff intervention, pharmacy, etc) and also a description of the care pathway workflow. In addition, instance level information is required to run one or possibly several care pathways simultaneously: the CD appointments and the services availabilities, according to a weekly pattern with some exceptions (e.g. public holidays).

Concerning fairness, basic ethical principles state that every patient deserves optimal care regardless of his medical condition or prognosis. This means that the scheduling must avoid degrading RDI. In the case an external event results in RDI degradation, more attention is paid to it to avoid further degradation. In case of resource shortage, this can become unachievable. However, the system should be able to detect and report such situations ahead of time to allow the management team to take corrective measures, like a transient increase of staffing or redirecting new patients to another unit. Note that we did not consider the nurse scheduling aspect which can be considered as independent and was covered by techniques presented in the can be first case study.

3.3 Designing for Fairness

Our approach is about scheduling the care of all patients together in such a way that some global time constraints are enforced. The actual situation in most day hospitals is that patients are scheduled on a first-come first-serve basis. With such a policy, in case of resource shortage (beds, nurses), the treatment of a patient might be postponed by some days. For some patients, such a delay can result in great harm in terms of chance of healing.

In contrast, our solution avoids resource shortage by smartly spreading over time the start date of the chemotherapy pathways. However, if resources were still limited, the system will smartly select patients to postpone by limiting the impact on their time constraints and thus their RDI.

The objective function to maximize is the global RDI over the pool of patients. We have developed two global criteria:

The first considered criterion was to **maximize the minimal RDI among the whole pool of patients**. It is implemented by minimizing the schedules makespan among all patients using *iFlatRelax* [24]. The schedule of a patient is an interleaving of appointments and resting periods, followed by a "stub" activity at the end. This stub is needed because all patients do not start their treatment at the same time. That stub activity enables us to consider their treatment duration instead of reasoning on their ending date. The implementation was carried out using the OscaR.CBLS engine by extending an available *iFlatRelax* for non-moveables tasks, forbidden zones and a more flexible model of resources. This criteria may look fair but patients with the highest "healing chances at start" (e.g. with no dose reduction) could be considered as "neglected".

Consequently, we considered a second criterion: **the maximisation of the summed RDI of all patients in the pool**. This can be modelled as a tardiness problem, i.e. overshot of a given point in time (patient dependent) multiplied by a constant. This problem is widely studied and was solved using a task swapping neighbourhood starting from a solution provided by *iFlatRelax* because it was tightly packed and computed very quickly.

3.4 Validating Fairness

Achieving a good level of validation of care pathways in a real day hospital setting environment is tricky not only because of organisational complexity but also because the collected data fall usually within a few standard scenarios. In order to understand the system under stressed conditions, we designed a complete simulation environment [34]. It has the ability to run over accelerated time. It is composed of a scenario driver able to execute scenario with specific profiles (e.g. adverse unexpected events, high load, etc). Event can also be injected through the standard web-based nurse interface. In addition, a KPI component records all important indicators such as RDI for all patients and the system load.

Several simulation sessions were organised together with oncology practitioners involving three hospitals (UCL/Cancer Institute, Grand Hospital of Charleroi and UZ Leuven). Typical simulation setting is a realistic unit of 10 bed capacity per day, i.e. 50 beds per week resulting in a 150 beds theoretical pathway capacity. We report here about the following two scenarios more specifically related to fairness:

– progressive load increase until reaching service saturation
– adverse serie of event systematically targeting the same subset of patients

Scenario 1 - Progressive Load Increase. The scenario is to progressively increase the load starting from an empty pathway until overflowing the unit capacity after about 40 weeks as a result of a greater number of patient entering the pathway than those leaving it. This capacity is about 80% due to the fact that treatment time is actually less than one day and the remaining time is not enough to fit an extra patient.

Figure 4 confirms the good stability of the system: the minimal RDI is kept constant as well as the foreseen load over the next weeks.

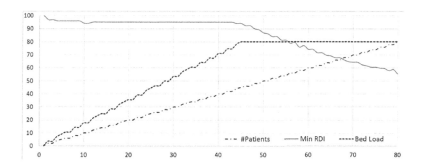

Fig. 4. Statistics for a long run simulation reaching saturation.

When looking at the variability in RDI across the patient pool, one can see the variability is quite contained and all RDI are kept above 90%. This variability usually result for single days of delay accumulating. Once the saturation is reached, however the system clearly shows it cannot cope anymore with RDI degrading below the acceptable 85% threshold and a wider variation (Fig. 5).

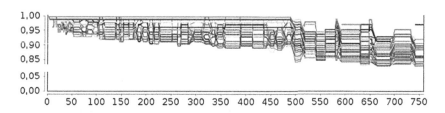

Fig. 5. KPIs for a long run simulation reaching saturation (about at tick 500 of the scheduler, each tick is a replanning event).

Scenario 2 - Adverse Events of a Defined Patient Subset. In this scenario, a patient is declared unavailable at the ideal delivery date and also at the three following days, as depicted in Fig. 6 showing the nurse interface.

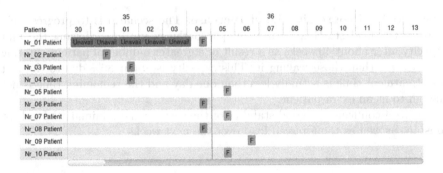

Fig. 6. User interface for managing patient constraints and appointments.

Figure 7 shows that in this case, the system will plan the patient directly after its unavailability period, resulting of course in an inevitable degradation of its KPI (about at tick 7). However other patient in the workflow are not impacted by this delay.

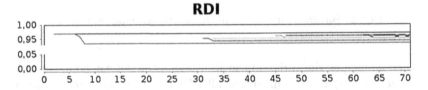

Fig. 7. Effect of a delay on the RDI of the patient pool.

4   Case Study 3 - Shared Shuttle System

4.1   Context

We consider here the problem of organising a shared pick-up and delivery service for the Sam-Drive company (or just "Sam" in short) located in the suburbs of Brussels in Belgium [8,25]. The Sam system has strong sustainability goals as mobility is a big challenge [1]. Before introducing the software, it was managed almost entirely manually.

However a number of key development goals required to consider introducing automated optimisation tools such as:

- *Decreasing operating costs* by minimising time on road (including traffic jams) and minimising driver idle time. This process will also decrease carbon emissions resulting in a positive environmental impact.

- *Empowering the dispatching team.* The previous situation was a poor IT solution (i.e. Excel sheets) putting stress on the team and also limiting the operational and growth capabilities.
- *Implementing more complex pricing models.* A key point is to encourage ride sharing with a decreasing pricing model for people accepting to travel with others and with some limited detour. This rewarding associated to this behaviour is a form of fairness.
- *Produce balance schedules for drivers.* The company use low-qualified labour, often unemployed people as drivers and offers them a secure job, in contrast to precarious jobs proposed by the "app-driven" new economy, like Uber [42]. Related to fairness, optimisation is better than humans to meet driver preferences and reach well-balanced schedules

4.2   Problem Statement and Fairness Requirements

Sam has two kind of clients, either for recurrent trips (typically for driving kids to/from school or extra-scholar activities) and "one-shot" trips (e.g. airport shuttle). As a result different optimisation functions are necessary: build a completely new schedule at the start of a semester, introduce extra trips in a schedule (mainly acceptance check) and reoptimise trips before producing the driver sheets.

For customers, the above two categories are not treated equally because there is a long term commitment for recurring drives. However this does not impact the quality of the schedule as it is managed before accepting a drive. Customers constraints are mainly time windows for departure and arrival, the former being generally more flexible. Customer may accepting a maximal detour. Its acceptable value is still evolving but its is expressed as an extra fraction of the direct drive time (like 50%) and possibly higher for short trips (e.g. 10 min for trips under 10 min). The rule is of course unique and the point is that people gets a discount proportional to the experienced detour time. Note recurring drives are carried out with the same passengers and usually the same driver (except in case of unavailability) because ruling out variability reduces risk of error/misunderstanding and gives more trust.

For drivers, the following requirements are taken into account for planning their schedules:

- working hours are defined for customers and are driver specific, e.g. some drivers accept to work at night while other no.
- for night work, a (legal) recovery time is enforced
- work time is based on unit of minimal 4 h (i.e. 4 h are paid even if there is only one drive). The scheduler must of course try to fill those minimal units as much as possible for the company profitability.

A resulting schedule will be more or less satisfied by its schedules based on different criteria such as how much it is paid for the time away from home, number of unpaid pauses (i.e. for slots exceeding 4 h), tight schedules with no time to come back home.

4.3   Designing for Fairness

For customers, fairness is achieved through the following techniques:

- time windows are always respected for all customers through the use of strong constraints
- the sharing behaviour is the default behaviour. The solver will decide on the amount of detour and the client will be rewarded according to the rule. Travelling alone is treated using an extra constraint. It is systematically applied for some time critical cases like going to an airport to catch a flight.
- global schedulability is (roughly speaking) achieved in two phases: recurring drives are planned two times a year (September, January: its is driven by the school agenda) and fills up the capacity at specific times (mainly start/end of day) leaving unallocated time for other drives (e.g. 9AM to 4PM). This drive time is then allocated incrementally based on incoming requests. Of course recurring drives can also evolve but in a smooth way.

For drivers, fairness among driver is not enforced by the engine itself given the drivers have very different profiles. Of course, legal constraints such as off time and rest time are enforced. Perceived fairness is tracked using specific KPI measuring the day "compactness", the presence or absence of unpaid pauses and other elements that can positively or negatively contribute to satisfaction. Those KPIs are available in a global dashboard for the dispatchers who can take specific preventive or corrective measures. At some point it could also become become part of the objective function.

4.4   Validating Fairness

This project is still in early validation phase and at this point the optimisation engine is only partly supporting the above fairness design and was not yet specifically validated for fairness. This also includes discussion with customers and drivers. For customers, surveys are currently ongoing to decide about the choice of the maximal detour a passenger is ready to accept and about the associated incentive. The idea is that the economic gain for sharing a ride is substantially redistributed to customers and not kept as an extra profit. So far this principle seems more important than the fine details of the redistribution itself.

5   Some Lessons Learned and Recommendations

5.1   Identifying Fairness Requirements in the Application Domain

As stressed in the introduction, the concept of fairness is difficult to define precisely. It should always be investigated within the scope of a given system or domain. Based on our experience, there a few entry points to identify fairness requirements and more importantly the need to address them.

- analysing the resource/task allocation process and related rules which may have legal roots (e.g. rest time for night work) or be company specific (e.g. permission to take break time at home for drivers). Starting from this, one should try to understand if it results in a specific advantage or disadvantage for the employee.
- the level of homogeneity of worker profile is also important to see if there is or not a competition for getting the advantage or avoiding disadvantage. In the Sam case, some drivers are interested by night work for the rest time or extra money they get while others will prefer to stay within working hours for health or family reasons. On the customer side, some can accept a detour while others not and get a reward on the ride price. In this case fairness is achieved through respecting preference as much as possible and/or enforcing compensation. In the case of competition among more homogeneous profiles (e.g. doctors having to perform a night shift once a month) then fairness should be more carefully enforced in the allocation process itself as described in the NiceWatch case. Other similar problems are scheduling shifted work in continuous production environment or for train drivers.
- the process might also express fairness requirements in explicit terms, e.g. patient equal access to care for ethical reasons. However this can translate in more complex technical requirements when considering the global organisation process like a clinical pathway because it needs to be combined with complex dose delivery requirements required for example for a chemotherapy.

5.2   Validating Fairness Through Quantitative Indicators

Given fairness requirements are quite fuzzy and that is it difficult to understand their impact, it is a good idea to try to quantify them through specific indicators. This is not only interesting to define them but it makes also possible to deploy some kind of monitoring before considering the need to take them into account in the scheduling process itself.

For example, in the Sam case, the driver satisfaction is still being defined and monitored. This process allows us to refine the definition in terms of positive and negative factor to consider and their relative importance/impact. In the case of NiceWatch, the indicators are integrated into the local search optimisation engine and also clearly reported at a very detailed level.

In some case, standard indicator might also exist and they are naturally adopted. For example, for cancer treatment the RDI has received been proposed and validated in medical journals.

5.3   Capturing Fairness in Local Search

A local search solution is composed of a model and a search procedure. Constraints to be optimised can be captured either as part of the state (stronger approach, not allowing violation) or has part of the objective function has preference and also allowing violations. Fairness is captured using the second mechanisms because they are not usually part of "core" constraints and also often added later in an improvement phase. Validated indicators easily translate into specific contribution for the objective function like the patient RDI or driver satisfaction.

In order to make the search more efficient, the search procedure might use specific neighbourhoods that will prioritise the search to favour fairness. For example, in the NiceWatch case, the affinity function is exactly that: it will select a doctor matching the preference but also favouring a deficit in fulfilling those preference in the past (to compensate on the long term). Although implemented in an ad-hoc way in this case, CBLS frameworks such as OscaR.CBLS offers facilities (invariants) to easily implement such selection in a very efficient way.

5.4   Evidencing Fairness

Claiming fairness is not enough: a system must be able to generate convincing evidence to people challenging it is achieved. In order to achieve this, one can either use a black box or white box techniques.

The *black box approach* is to just look at the result and check the claimed indicator without looking inside the optimisation process. This also require to monitor on a longer period if fairness is achieved over time (e.g. a few month for night watches). It does also not provide evidence the process will continue to behave as expected.

The *white box approach* is to be able to provide full transparency of the optimisation process itself. The process should provide an explanation of how it took a decision according to well defined rules and that is was not manipulated. This approach was taken for NiceWatch and also was one of the motivation for using an ad-hoc solver. While a off-the self solution could achieve higher quality results (i.e. less discomfort) than our approach and also cope with fairness, it is hard to achieve a good level of transparency with them and hence there is a risk of early rejection. Our approach on the contrary is able to achieve transparency about "even discomfort". It has also the capability to evolve to reduce the level of discomfort. In the end, the overhead of having to implement the algorithms without relying on a framework is also not so high when balanced with those advantages.

Other important related points are:

- in order to be easy to understand, the rules should not be overly complex, otherwise the explanation will be difficult to understand and people will not trust the system to be fair.
- the explanations generated by the solver should processed for maximising their understanding by the users: raw "listing" should be avoided and graphics should be favoured over tables aligning numbers.
- execution should be deterministic to make sure it is not altered by repeated runs.

5.5   Anticipating Fairness Issues

In specific cases, an on-line interface might help in critical process which may alter fairness. An example for care pathway is to fix the next appointment date for a patient with some unavailability: the system can provide a direct feedback on a safe date range for the patient and not altering other schedules. On the day clinic side, resource availability should also be secured (nurses, rooms) and checked as early as possible in specific application (e.g. holiday and cleaning/maintenance). Another example for the Sam system is that an incoming request is first examined for feasibility before being accepted and fully optimised.

6   Related Work

In this section, we review experiences reported by other in the same domains as our case study (hospital, routing) but also other domains like airlines, communication networks, education. We also consider other optimisation techniques than local search.

6.1   Doctor and Nurse Scheduling Problems

In DSP, fairness constraints are identified along other constraints and typically formulated as the fair distribution of different types of shifts among doctors with the same experience in [14]. Fairness received specific attention in the emergency room context [10,11,36].

MIP based heuristics have been used to create balanced scheduling from the set of doctors [10]. Integer programming has also been used to take into account constraints of the schedule, different preference ranks w.r.t. shifts, and the historical data of previous schedule periods to maximize the global satisfaction about the proposed shift schedule [18]. The resulting shifts and days-off were fair and met the staff satisfaction.

In local search, an objective function is expressed as a weighted sum of soft constraint violations. Such an objective function has the advantage of being both easy to understand and to implement. However, they can produce unfair solutions because some high quality allocations can compensated low quality ones. A solution proposed by [37] is to use a function where the quality of the worst individual allocation will directly impact the overall solution quality. In doing so, a planing will not be improved at the expense of the worst individual case. Experimental results have confirmed the resulting solution is more fair, nevertheless a drawback is that the search seems less efficient given the new structure of the function. In addition to the lack of explanation traceability, this reinforces us about our dedicated approach.

A complete overview of techniques for NSP with some hint about how to come with personal constraints is presented in [5]. Evolutionary algorithms are quite commonly used and an approach for the formulation of the fitness function has proven to be very powerful both to enable extendibility and to provide a quick and explanatory mechanism. We achieved the same results using our own approach and our belief is that the technique used is not the key point but rather the ability to take into account the right set of constraints, including historical data as well as the ability to produce justifications. The requirement for traceability also favour better architecture which in turn ease the ability to deal with more complex real-world constraints.

6.2   Fairness Indicator and Strategies

A number of fairness indicators (or metrics) have been defined, often in the context of economics for measuring the distribution of wealth or more technically in computer networks for solving the problem of bandwidth allocation.

- the *Max-Min Fairness* (MMF) is achieved if trying to "favour" (i.e. allocate more resource) someone can only be achieved by "defavouring" someone else. As a result, resources are optimised by raising priority to smaller demands [3].
- the *Jain Fairness Indicator* (JFI) is a way to rate fairness of allocation between n users. It maximal value is 1 is reached upon equal allocation while its worst value $(1/n)$ means a single user gets all the resource [16].
- the *Generalized Gini Indicator* (GGI) is a well-know inequality measure defined in economics and used both for fairness and Pareto-efficiency [26,41]. It is a weighted sum of function measuring the deviation from the point of equality for each element. It is suitable to use in multi-objective optimisation because it can achieved a balanced cost vector [6].

Such indicators have also been used in other domains. For example, the curriculum-based course timetabling (CB-CTT) considers the problem of creating fair course timetables in an university context. To manage fairness, a key idea is that violations of soft constraints in the produced timetables, should be distributed in a fair way among the stakeholders. The above approaches based of max–min fairness and Jain's fairness index have both been studies [23]. In aircraft landing, the fairness can be measured through comparing against the initial schedule either on a relative scale or as absolute deviation. The fairness is then achieved by minimising the maximal deviation, which has similarities with max-min approach and GGI.

In our case studies we have not explicitly used those indicators. However our initial strategy for fair scheduling of care pathway was actually close to a max-min criteria over the patient RDI. For night watches, we use a domain specific affinity indicator which includes a fairness component. At this point with have not used multi-objective techniques and not considered the GGI.

7 Conclusion

In the paper, we investigated about how to cope with fairness while optimising systems. We took a practical approach based on the analysis of three cases studies in different domains. We analysed how fairness requirements could be identified, modelled, implemented and validated, in the specific local search context. By putting together those cases and looking at related work, we could produce a set of recommendations aiming at providing some guidance across the full lifecycle from capturing fairness within a specific domain to presenting how they are addressed in an effective way. While not exhaustive, we believe our work can help practitioners having to address fairness requirements. For example, we got interesting feedback about the importance of perceived fairness over the level of quality of a solution, which motivated the choice to design a less efficient but highly adaptable and traceable solution for the NiceWatch case [33].

As future work, we plan to keep elaborating a more structured and complete knowledge body for this important kind of constraint, possibly also attracting other contributors to this effort. A first task is to enrich our guidelines by analysing more cases from our own experience and from the literature (with some of them already highlighted in our related work section). Based on this, we also plan to provide a more complete taxonomy of fairness from the requirements engineering point of view. Finally, we also plan to consider a wider range of implementation techniques than the current scope limited to local search.

Acknowledgements. This research was partly funded by the Walloon region as part of the PRIMa-Q CORNET project (nr. 1610019). We warmly thanks MedErgo and Sam-Drive for allowing us to share their respective cases.

References

1. Banister, D.: The sustainable mobility paradigm. Transp. Policy **15**(2), 73–80 (2008)
2. Benoist, T., Estellon, B., Gardi, F., Megel, R., Nouioua, K.: Localsolver 1.x: a black-box local-search solver for 0–1 programming. 4OR **9**(3), 299–316 (2011)
3. Bertsekas, D., Gallager, R.: Data Networks. Prentice-Hall, Upper Saddle River (1987)
4. Boulware, L.E., Troll, M.U., Wang, N., Powe, N.R.: Perceived transparency and fairness of the organ allocation system and willingness to donate organs: a national study. Am. J. Transplant. **7**(7), 1778–1787 (2007)
5. Burke, E.K., et al.: Fitness evaluation for nurse scheduling problems. In: Proceedings of the IEEE Congress on Evolutionary Computation, vol. 2, pp. 1139–1146 (2001)
6. Busa-Fekete, R., Szörényi, B., Weng, P., Mannor, S.: Multi-objective bandits: Optimizing the generalized gini index. CoRR abs/1706.04933 (2017). http://arxiv.org/abs/1706.04933
7. Campbell, H., Hotchkiss, R., Bradshaw, N., Porteous, M.: Integrated care pathways. Br. Med. J. **316**, 133–137 (1998)

8. CETIC and Sam-Drive: Samobi - the next generation shared taxi (2016). https://www.cetic.be/SAMOBI-3055

9. van Dam, P.A., et al.: A dynamic clinical pathway for the treatment of patients with early breast cancer is a tool for better cancer care: implementation and prospective analysis between 2002–2010. World J. Surg. Oncol. **11**(1), 70 (2013)

10. Devesse, V., Santos, M.O., Toledo, C.: Fairness in physician scheduling problem in emergency rooms. In: Revista de Sistemas de Informação da FSMA, pp. 9–20 (2016)

11. Ferrand, Y., et al.: Building cyclic schedules for emergency department physicians. Interfaces **41**(6), 521–533 (2011)

12. Francez, N.: Fairness. Texts and Monographs in Computer Science, 1st edn. Springer, New York (1986). https://doi.org/10.1007/978-1-4612-4886-6

13. Gecode Team: Gecode - an open, free, efficient constraint solving toolkit (2017), available under the MIT licence from http://www.gecode.org/

14. Gendreau, M., et al.: Physician Scheduling in Emergency Rooms. In: Burke, E.K., Rudová, H. (eds.) PATAT 2006. LNCS, vol. 3867, pp. 53–66. Springer, Heidelberg (2007). https://doi.org/10.1007/978-3-540-77345-0_4. http://dl.acm.org/citation.cfm?id=1782534.1782540

15. Gooch, P., Roudsari, A.: Computerization of workflows, guidelines, and care pathways: a review of implementation challenges for process-oriented health information systems. J. Am. Med. Inform. Assoc. **18**(6), 738–748 (2011)

16. Jain, R., Chiu, D., Hawe, W.: A quantitative measure of fairness and discrimination for resource allocation in shared computer systems. CoRR cs.NI/9809099 (1998). http://arxiv.org/abs/cs.NI/9809099

17. Landtsheer, R.D., Delannay, G., Ponsard, C.: Dealing with perceived fairness when planning doctor shifts in hospitals. In: Proceedings of the 7th International Conference on Operations Research and Enterprise Systems, ICORES 2018, Funchal, Madeira - Portugal, pp. 320–326, 24–26 January 2018

18. Lin, C.C., Kang, J.R., Liu, W.Y., Deng, D.J.: Modelling a Nurse Shift Schedule with Multiple Preference Ranks for Shifts and Days-Off. Mathematical Problems in Engineering (2014)

19. Lyman, G.: Impact of chemotherapy dose intensity on cancer patient outcomes. J. Nat. Comput. Canc. Netw. **7**, 99–108 (2009)

20. Marynissen, J., Demeulemeester, E.: Literature review on integrated hospital scheduling problems. Tech. Rep. 555258, KU Leuven, Faculty of Economics and Business (2016)

21. McGlynn, E.A., et al.: The quality of health care delivered to adults in the United States. N. Engl. J. Med. **348**(26), 2635–2645 (2003)

22. MedErgo: NiceWatch - Complex Schedules within Seconds (2016). http://www.nicewatch.net

23. Mühlenthaler, M., Wanka, R.: Fairness in academic course timetabling. Ann. Oper. Res. **239**(1), 171–188 (2016). https://EconPapers.repec.org/RePEc:spr:annopr:v:239:y:2016:i:1:d:10.1007_s10479-014-1553-2

24. Michel, L., Hentenryck, P.V.: Iterative relaxations for iterative flattening in cumulative scheduling. In: Proceedings of 14th International Conference on Automated Planning & Scheduling (ICAPS) (2004)

25. Michel, R.: Sams (2012). https://www.sam-drive.be

26. Moulin, H.: Fair Division and Collective Welfare. MIT Press, Cambridge (2003). http://eprints.gla.ac.uk/86973/

27. NHS: Good practice guide: Rostering (2016). https://improvement.nhs.uk/uploads/documents/Rostering_Good_Practice_Guidance_Final_v2.pdf

28. NSW: Principles of rostering (2015). http://www.health.nsw.gov.au/Performance/rostering/Pages/principles.aspx
29. Ogryczak, W., Luss, H., Pioro, M., Nace, D., Tomaszewski, A.: Fair optimization and networks: a survey. J. Appl. Math. **2014**, 25 (2014)
30. OR-tools Team: OR-tools: Operations research tools developed at Google (2017). https://code.google.com/p/or-tools/
31. OscaR Team: OscaR: Operational Research in Scala (2012). available under the LGPL licence from https://bitbucket.org/oscarlib/oscar
32. Piccart, M., Biganzoli, L., Di Leo, A.: The impact of chemotherapy dose density and dose intensity on breast cancer outcome: what have we learned? Eur. J. Can. **36**(Suppl 1), 4–10 (2000)
33. Ponsard, C., Landtsheer, R.D., Germeau, F.: Building sustainable software for sustainable systems: case study of a shared pick-up and delivery service. In: Proceedings of the 6th International Workshop on Green and Sustainable Software (accepted), GREENS@ICSE 2017, Gothenburg, Sweden, 27 May 2018
34. Ponsard, C., Landtsheer, R.D., Guyot, Y., Roucoux, F., Lambeau, B.: Decision making support in the scheduling of chemotherapy coping with quality of care, resources and ethical constraints. In: ICEIS 2017 - Proceedings of the 19th International Conference on Enterprise Information Systems, Porto, Portugal, 26–29 April 2017
35. Roucoux, F., et al.: Pipas - optimal piloting of care pathways. Final Report, Université catholique de Louvain (2014)
36. Santos, M., Eriksson, H.: Insights into physician scheduling: a case study of public hospital departments in Sweden. Int. J. Health Care Qual. Assur. **27**(2), 76–90 (2014). MCB University Press
37. Smet, P., Martin, S., Ouelhadj, D., Ozcan, E., Berghe, G.V.: Investigation of fairness measures for nurse rostering. In: Practice and Theory of Automated Timetabling (PATAT), Son, Norway (2012)
38. Stutzle, T.: Local Search Algorithms for Combinatorial Problems: Analysis, Improvements, and New Applications. Ph.D. Thesis, Infix Verlag (1999)
39. Van Hentenryck, P., Michel, L.: Constraint-Based Local Search. MIT Press, Cambridge (2009)
40. Walter, F., et al.: Success of clinical pathways for total joint arthroplasty in a community hospital. Clin. Orthop. Relat. Res. **457**, 133–137 (2007)
41. Weymark, J.A.: Generalized GINI inequality indices. Math. Soc. Sci. **1**(4), 409–430 (1981)
42. Younglai, R.: Rise of sharing services Uber, Airbnb points to a precarious labour climate. The Globe and Mail (2015). http://bit.do/precarious-sharing-economy

Author Index

Printed in the United States
By Bookmasters